43rd Annual Meeting of the Plant Growth Regulation Society of America 2016

Raleigh, North Carolina, USA
17 – 21 July 2016

ISBN: 978-1-5108-4499-5

Printed from e-media with permission by:

Curran Associates, Inc.
57 Morehouse Lane
Red Hook, NY 12571

Some format issues inherent in the e-media version may also appear in this print version.

Printed by Curran Associates, Inc. (2017)

Plant Growth and Regulation Society of America
1018 Duke Street
Alexandria, VA 22314
USA

Phone: 703-836-4606 x116
Fax: 703-836-4607

cslone@ashs.org

Additional copies of this publication are available from:

Curran Associates, Inc.
57 Morehouse Lane
Red Hook, NY 12571 USA
Phone: 845-758-0400
Fax: 845-758-2633
Email: curran@proceedings.com
Web: www.proceedings.com

MEETING SUMMARY

43rd Annual Meeting of the Plant Growth Regulation Society of America

Chris Gunter
Meeting Convener & 1st Vice President, PGRSA

The 43rd annual meeting of the Plant Growth Regulation Society of America was held in exciting downtown Raleigh, North Carolina. PGRSA members were introduced to the sights, sounds and tastes of downtown Raleigh by opening the meeting with a preconference food tour. Attendees sipped, tasted, walked and talked, getting to know each other and their new temporary home in the heart of North Carolina's capital city. The tour's final stop included a personal introduction to the state's Museum of Natural Sciences with plenty of time to explore on their own.

The meeting featured a wonderful lineup of keynote speakers and contributed presentations, including Bernard Glick, from the University of Waterloo; Becky Langer, from Bayer's Bee Care Center; Kevin Folta, from the University of Florida; Daniel Leskovar, from Texas A&M University; Raquel Hontecillas-Maragazo from the Biocomplexity Institute of Virginia Tech; and David Hiltz, of Acadian Seaplants. Our conference was kicked off with a wonderful Historic Perspectives talk delivered by Cary Mitchell of Purdue University. In addition, our Young Scientist of the year award winner, Maria Pozo, from the Estacion Experimental del Zaidin, CSIC, presented her research.

This year's meeting attracted six students to participate in the poster and oral presentation competitions, and four student travel awards were given this year. The two winners were Joshua B. Henry, North Carolina State University, who presented the top student poster entitled, "Phosphorus Restriction As An Alternative Method Of Growth Control For Petunia," and Bizhen Hu, The Ohio State University, who presented the top student oral paper entitled, "Separate and Interactive Effects of Light Intensity and Temperature on Healing of Grafted Tomato Seedlings."

The program committee members for the meeting included Casey Barickman, Gurreet Brar, Chris Gunter, Holly Little, Sonja Maki, Carl Sams ,Tim Spann, Jozsef Racsko, and Michael Retshwisch. All of these members are to be commended for their hard work and creativity in putting together a fantastic program.

On behalf of the PGRSA Steering Committee, I would like to invite anyone who may have an interest in plant growth regulation to attend the 44th Annual PGRSA meeting to be held in Anchorage, Alaska, August 6-10, 2017! We look forward to seeing you there.

HONORARY LIFETIME MEMBERS

Anson Cook
Aldo J. Crovetti
Eric A. Curry
Horace G. Cutler
Richard Dunand
C. Dave Fritz
Masayuki Katsumi (JSCRP)
Wayne Mackay
Wilhelm Rademacher
Edward F. Sullivan
Thomas Tworkoski

Acknowledgements

The Proceedings has been published yearly since the fourth annual meeting in 1977 at no charge to the author(s). Papers in this edition were received electronically and, although we have made minor changes in formatting, every effort has been made to preserve the original submissions, the content of which remains the sole responsibility of the respective author(s).

The Proceedings were compiled by:
Dr. Chris Gunter, 1st Vice President, Meeting Organizer
Horticultural Science Department
North Carolina State University
230 Kilgore Hall, Box 7609
Raleigh, NC 27619-7609
Office 919-513-2807
Cell 919-368-2724
Fax 919-515-2505
chris_gunter@ncsu.edu

The Proceedings are available for download at www.PGRSA.org. The most recent two years of the Proceedings are available to PGRSA members only. Older issues are available to everyone free of charge.

Hard copies of the Proceedings are available for a fee from Curran Associates, Inc. print on demand services. To order a hard copy visit http://www.proceedings.com/.

A SPECIAL THANK YOU TO THE FOLLOWING COMPANIES FOR FINANCIAL SUPPORT OF THE MEETING

Platinum Sponsor

VALENT BIOSCIENCES CORPORATION

Gold Sponsors

fine

Mycorrhizal Applications

Silver Sponsors

acadian

Ocean Organics

Bronze Sponsors

LumiGrow
The power of spectrum control.

REPAR

ohp Partners with solutions

2015 – 2016 PGRSA Steering Committee

Dr. Holly Little, President
Acadian Seaplants Ltd.
2443 Fair Oaks Blvd., #67
Sacramento, CA 95825
Office 530-219-5415
hlittle@acadian.ca

Dr. Carl Sams, Past President
Institute of Agriculture – Plant Sciences Department
2431 Joe Johnson Drive
Knoxville, TN, 37966-4561
Office 865-974-8818
Fax 865-974-1947
carlsams@utk.edu

Dr. Chris Gunter, 1st Vice President
Horticultural Science, North Carolina State University
230 Kilgore Hall, Box 7609
Raleigh, NC 27619-7609
Office 919-513-2807
Cell 919-368-2724
Fax 919-515-2505
chris_gunter@ncsu.edu

Dr. Jozsef Racsko, 2nd Vice President
Valent Biosciences Corp.
870 Technology Way
Libertyville, IL 60048
Office 847-968-4720
Cell 224-456-1366
jracs@valent.com

Mr. Michael Rethwisch, Secretary
University of Nebraska – Lincoln
451 N. 5th Street
David City, NE 68632
Office 402-367-7410
mrethwisch2@unl.edu

Dr. Nicole Waterland, MAL-1
Division of Plant and Soil Sciences
West Virginia University
Morgantown, WV 26506-6108
(304) 293-2969
Nicole.Waterland@mail.wvu.edu

Dr. Casey Barickman, MAL-2
North Mississippi Research and Extension Center
Mississippi State University
PO Box 1690
Verona, MS 38879
Office 662-566-8001
tbarickm@ext.msstate.edu

Dr. Gurreet Brar, MAL-3
Assistant Professor/Citrus Horticulture Extension Specialist
University of Florida/IFAS/SWFREC
2685 SR 29 N, Immokalee FL 34142 USA
239-658-3426 (office)
gbrar@ufl.edu

Dr. Tim Spann, Business Manager
California Avocado Commission
12 Mauchly, Suite L
Irvine, CA 92618
Office 949-341-1955
Cell 863-258-5950
tim@pgrsa.org

Dr. Sonja L. Maki, Executive Director
Biology Department, Carleton College
One North College Street
Northfield, MN 55057
Cell 715-690-4334
Fax 507-222-5757
smaki@carleton.edu

Ms. Cindy Slone, Executive Secretary
PGRSA/American Society for Horticultural Science
1018 Duke Street
Alexandria, VA 22314
Office 703-836-4606
Fax 703-836-4607
cslone@ashs.org

TABLE OF CONTENTS

LED LIGHTING AS A PLANT-GROWTH REGULATOR: AN HISTORICAL PERSPECTIVE

Cary A. Mitchell
Department of Horticulture & Landscape Architecture Purdue University, West Lafayette, IN 47907-2010
E-mail: cmitchel@purdue.edu

The first practical visible-spectrum (red) light-emitting diode (LED) was developed in 1962 by General Electric Engineer Nick Holonynak from a semi-conductor alloy of gallium arsenide phosphide (GaAsP). The first LEDs were of very low power and were used as indicators in electronic equipment such as alarm-clock radios. As LEDs gained incrementally in power through engineering advances, their utilization expanded to traffic lights and auto brake lights. LEDs were not powerful enough to provide adequate light for plant growth and development until the 1980s. So-called "high-output" LEDs today put out from one to several watts of photon power individually. Because of their small individual size, massive arrays of closely spaced discrete LEDs can put out substantial photon power, but also substantial waste heat, requiring active heat sinking to preserve photon-emission efficiency and lifespan.

LEDs are solid-state light- emitting devices constructed from different semiconductor materials with a junction between them (Fig. 1, Refs. 1& 2). Applying an electrical current to the diode causes atoms in both semiconductor materials to flow toward the junction. When they get close to each other, n-type atoms having extra electrons donate them to the p-type atoms, which accept them. As n electrons fall into p "electron holes", extra energy is released in the form of photons. The wavelength or color of the photons released depends on the chemical composition of the different semi- conductor materials making up the diode. This type of light production is known as "electroluminescence". Because there are no electrodes or filaments, the "electronic nature" of LED photon emissions gives them the potential for very long lifespans, provided that they are not allowed to operate at elevated temperatures for prolonged periods of time.

Many "discrete" LEDs are combined to form an "array" that can put out a solid beam of light (Fig. 2). Otherwise, their output is a "point source" of light whose intensity decays exponentially with the square of linear distance between the LED and what is being lighted. A discrete LED in an alarm clock might be 0.1 Watt. An LED in a strand of Christmas tree lights would be less than that.

LEDs are "solid state", which means they are rugged, not fragile. They do not take up a lot of space individually and they are not heavy. They have a long lifespan. By manipulating semi-conductor composition, the color of light given off can be controlled. They operate at low-voltage direct current. Waste heat can be "heat-sinked" out of the back of the diode, so photon-emitting surfaces are relatively cool. Because of that, energized LEDs can be located fairly close to plant tissues, and desired photon flux can be achieved at low current flow, without loss of intensity due to large separation space otherwise required by light sources that give off copious radiant (waste) heat. The relative "coolness" of LED photon-emitting surfaces helps negate effects of the inverse square law ($I \alpha d-2$).

For all of those reasons, LEDs were considered a promising source of light for plant growth in space, so NASA space life science programs sponsored the first plant research with LEDs. The first LED/plant-growth paper was authored by Ray Bula, Bob Morrow, Ted Tibbitts, and Dan Barta at the University of Wisconsin, as well as others from industry (Fig. 3, Ref. 3). The red LEDs first used were composed of gallium-aluminum arsenide (GaAlAs) that gave an emission peak at 660 nm. The red LED light plus 30 μmol/m2/s of blue fluoresecent light (because blue LEDs were not developed enough at the time) for a total PPF of 325 μmol/m2/s caused 'Grand Rapids' lettuce plants to grow equivalent to that under the same intensity of cool-white fluorescent and incandescent light sources, but at significantly lower input energy. To find the minimum amount of blue light needed, the Wisconsin group measured lettuce-seedling hypocotyl and cotyledon extension, which are excessive in the presence of red light alone (Fig. 4, Ref. 4). More than 15 μmol/m2/s blue was required to prevent excessive elongation.

The relative quantum efficiency (RQE) of photosynthesis as a function of wavelength indicates that red light is most efficient, with a secondary peak in the blue, and lower but still substantial efficacy of green by whole plants (5, 6). All wavelengths from 400 to 700 nm drive photosynthesis with varying efficiencies. Red LEDs typically are the LED wavelength of choice for growing plants, with other colors added as needed.

Researchers in Japan during the early 1990s, most notably Kozai, tested LEDs as sources of red and far-red light to control morphology of potato plantlets growing in sterile culture for micro-propagation purposes (Fig. 5, ref.7). FR light was found to promote stem elongation just as blue light represses it.

In both the U.S. and Japan during the 1990s, researchers tested combinations of red, blue, and far-red LEDs for effects on photosynthesis for a wide variety of plant species (8, 9, 10, 11). Blue together with red usually stimulated photosynthesis more than did red alone, and FR had indirect effects on photosynthesis by having direct effects on photomorphogenesis. This proved that you don't absolutely need broad-band light to drive photosynthesis and plant growth, which can have important implications for saving energy by avoiding less efficient wavelengths of PAR.

Researchers at the Wisconsin Center for Space Automation and Robotics (WCSAR), developed LED flight hardware for growing plants in space (12, 13, 14). Two grad students from Ted Tibbitts lab, Dan Barta and Bob Morrow, went on to develop careers in the space-plant growth-LED arena, Dan going with NASA at the Johnson Space Center, and Bob initially with WCSAR, and now with Orbital Technologies (ORBITEC), recently acquired by Sierra Nevada. Different generations of plant flight hardware have been developed, and Bob has been involved with much of it. WCSAR's Astroculture unit was used to grow wheat and dwarf rapid-cycling brassica in space. Croxdale et al. (15) and Stankovich et al. (16) reported successful formation of aerial potato tubers in space and seed-to-seed growth of Arabidopsis on ISS, respectively, both using flight hardware lighted with LEDs. Ground testing of plants grown under red plus blue LEDs for space hardware indicated excellent response to 325 µmol/m2/s of PAR with 30 µmoles of PPF coming from blue fluorescent lamps behind the transparent plastic LED mounts (Fig. 6). By the time the unit flew in space, blue LEDs had become efficient enough to include in the LED array per se. Wheat and brassica flew in Astroculture units, but lettuce didn't fly until ORBITEC launched its Veggie unit in 2014.

The Kennedy Space Center (KSC) became a focal point of plant research for space biology and bioregenerative life support in the 1990s (17, 18, 19). Civil servants such as John Sager and later Ray Wheeler monitored university grants involving LEDs and developed their own core contractor research group at KSC. Their research was basic in approach but mission oriented toward real spaceflight conditions. They were interested in how productive plants could be under different wavelengths from LEDs as sole-source lighting, and why.

Solanaceous plant species in particular growing under narrow-spectrum or low- intensity white light may develop a tumor-like physiological disorder called "intumescence" (20, 21). Bob Morrow did his PhD dissertation research on this disorder, finding it to be caused by lack of UV radiation, but also sometimes preventable by increasing FR light, suggesting phytochrome involvement. When Gioia Massa grew cowpea under sole-source R + B LED lighting with blue 5% of total PPF, intumescences developed on leaves and shoot tips, causing a decline in productivity (Fig. 7). If blue was increased to 15% of PPF, cowpea did not develop intumescence. However, increasing blue light to that level or above does not prevent intumescence for solanaceous species that have been tested. Another disorder that can happen with narrow-band LED lighting is "photobleaching" of leaf tissue when plants grow up against an LED (Fig. 8). It isn't thermal scorching, and it can happen even if photon flux is less than white-light saturation for photosynthesis. The plastic lens around individual LEDs may focus the light beam so narrowly that pigments can't handle the concentrated beam of photons at close quarters, and red light, with the highest RQE of any PAR wavelength, may oversaturate chlorophyll pigments.

In a KSC study comparing effects of different light qualities on growth and development of leafy greens (22), red, blue, and a moderate amount of green light (24%) stimulated leaf expansion and shoot dry-weight gain compared to everything else, but if a lot of green (86%) was given, productivity was inhibited (Fig. 9). Since green is absorbed less efficiently than red and blue, the green component of the RGB treatment may penetrate more deeply into the self-shading leaf canopy, driving photosynthesis in otherwise mutually shaded lower leaves. Where most of the light was green, productivity was low, likely because the green was absorbed inefficiently. Kevin Folta suggests that the high-green treatment also interfered with photomorphogenic actions of the limited red and blue present, inhibiting expansion of photosynthetic surface (23). "Green light, a signal to slow down or stop."

In 2007, members of the ASHS Controlled Environments Working Group sponsored a workshop on "LEDs in Horticulture". The event attracted much interest from the plant-research community, and publication of the proceedings in December 2008 attracted even more interest (Fig. 10). It has been one of the most highly cited issues in the ASHS literature. NASA Civil Servant John Sager is shown on the HortScience cover working with ORBITEC's LED-lighted Veggie Units.

Veggie was launched to the International Space Station (ISS) in 2014, where astronauts presently are growing vegetables on orbit (Fig. 11). Veggie has red, green, and blue LEDs, and its transparent accordion walls are upwardly expandable as plants grow in height. Astronauts wear sunglasses when working around the concentrated beams of LED light, as anyone should.

The HortScience article "Plant Productivity in response to LED Lighting" included in the 2008 LED workshop proceedings (24) has been the most highly read and cited ASHS article for most months over the past 8 years. It addresses many issues mentioned in this historical perspective.

In addition to identifying spectral requirements for productive growth of plants for space life support as well as for terrestrial controlled-environment agriculture, the next question relates to how the unique properties of LEDs might be leveraged to deliver light where it is needed while maximizing energy savings. Cary Mitchell's laboratory at Purdue University initially addressed the energy challenge by establishing proof of concept that "intracanopy lighting" of closed crop stands with low-wattage fluorescent lamps could save significant energy while improving crop productivity (25, 26, 27).

Because of the "hot" photon-emitting surfaces of traditional light sources and being aware of the relatively cool photon-emitting surfaces of LEDs, Mitchell developed the "lightsicle" concept for intracanopy lighting with sequential bottom-up on- switching of vertical arrays of LEDs as closed crop stands grow in height (Fig. 12). This allows lower electrical power to be used to deliver adequate PPF with close LED-plant spacing, thereby avoiding disadvantages of the inverse square law of point-source radiation with increased separation distance.

Working with Bob Morrow and Jeff Emmery at ORBITEC, the Mitchell lab tested first-generation lightsicles populated with 1-inch-square "light engines", each containing 100 small printed-circuit LEDs (Fig. 13). Plant tissues could "touch" a light engine and still receive a uniform blend of light colors without thermal scorching. The aluminum-alloy bar mounting strips were hollow and provided a vertical pathway for convectively removing waste heat from behind each diode with a fan drawing air up through the slotted bar.

In studies comparing intracanopy vs. overhead LED lighting of cowpea crop stands growing hydroponically in separate growth compartments of a controlled-environment room, Gioia Massa found much more leaf litter accumulation in the overhead-lighted stand, indicating limited light levels in the lower canopy (Fig. 14). More plant biomass accumulated in the intracanopy-lighted stand at significantly lower energy cost due in part to sequential switching-on of light engines from the bottom up as the crop stand grew in height.

The lightsicles also were designed to sense the location and size of plant tissue at different distances from the LED strip, and to selectively switch on LEDs to minimize waste of electrical energy by not lighting empty spaces (Fig. 15).

Working with Mike Bourget and Bob Morrow at ORBITEC, Lucie Poulet in the Mitchell lab next tested close-canopy overhead lighting of low-stature rosette crops such as leaf lettuce (28). Once again, close placement of LEDs to plants helped minimize energy expenditure compared to traditional lamps requiring substantial separation distance. These arrays also allowed selective switching of LED clusters to light only spaces below the array populated by plant tissues. Such "targeted" LED lighting saved another 50% in light energy (Fig. 16). Still more optimization of such smart lighting systems is possible and needed.

In partial spinoff from the space program, a multi- institutional group of academic scientists, economists, and industry engineers was awarded a USDA grant entitled "Developing Light- Emitting Diode (LED) Technology and Practices for Sustainable Specialty-Crop Production" (NIFA SCRI grant 2010-51181-21369). Participants in that SCRI project looked at developing guidelines for comparing LED technology with traditional lighting technology, LED engineering, economics of adopting disruptive LED lighting by the industry, propagation of vegetable crops with LEDs, propagation of bedding plant crops with LEDs, supplemental lighting of high-wire tomato year-round in greenhouses, developing novel LED technology, and floral induction and development for different photoperiodic classes of ornamental crops using LEDs. It would be impossible to summarize all of the accomplishments of that productive and diverse group in the remaining available space. The culmination of that project included a jointly written Horticultural Reviews article citing about 250 references on LEDs and plant growth (Fig. 17, ref. 29).

The research approach of the SCRI project emphasized supplemental lighting applications in the greenhouse. The Mitchell lab extended its NASA- sponsored intracanopy lighting approaches from small and medium-statured vegetable crops in growth chambers to high-wire tomato crops in the greenhouse. "LED towers" developed in collaboration with ORBITEC were extended versions of the original lightsicles. They use vertical strips of discrete high-output, dimmable red, blue, and far-red LEDs, are self-monitoring for power and energy consumption, and LED panels can be energized on both sides of a tower. The towers also can be switched off from the bottom up as tomato crops are harvested from the bottom up and vines are lowered and leaned (Fig. 18).

There have been two significant outcomes of this research: one is that fruit yield is boosted by supplemental lighting per se, compared to unsupplemented controls (Fig. 19). LED intracanopy lighting was no more effective than overhead HPS lighting, or a hybrid of the two, in terms of stimulating yield. However, LED intracanopy lighting gave considerable energy savings compared to overhead HPS, regardless of season. This was due to a combination of factors, including the staged switching on and off of selected panels on LED towers, plus lower power needed to achieve target PPF.

PhD graduate student Celina Gomez did the LED-tomato productivity studies. Articles published from that study include a report of LED tower capabilities and energy-savings from initial experiments (30). Another article included results from an entire year of experiments, one going from winter to summer when solar DLI was increasing, and the other from summer to winter with the solar contribution decreasing (31). A third article compared effects of different spectra of supplemental lighting at different times of year for propagating tomato transplants (32).

A fourth article characterized physiological responses of leaves at different positions along the downward vertical profile of high-wire tomato (33). In that physiology study, photosynthetic rate of leaves was measured as a function of light intensity (Fig. 20). The left column depicts light-saturation curves from a winter-to-summer experiment, on the right from summer- to-winter. The top pair of curves represents upper leaves of a vertical canopy, the center two panels the middle leaves, and the bottom pair lower leaves. Four supplemental lighting treatments were compared. For both seasonal-transition experiments, effects of leaf position in the canopy become increasingly significant with vertically downward leaf position in terms of leaf potential to respond to supplemental light. Controls always had the least capability to respond to increasing light, but position in the canopy still played a role, regardless of whether supplemental lighting was used. Summer-to-winter clearly had reduced photosynthetic potential compared to winter-to-summer.

Another question of significant importance regards whether LED lighting can improve quality attributes of plants. In a study conducted by Gary Stutte at KSC (Fig. 21, Ref. 34), lettuce that is naturally red outdoors was mostly green indoors under sole- source LED lighting with little or no blue light. However, the same lettuce accumulated increasing amounts of anthocyanin as the dosage of blue light used increased in proportion with red. The numbers in yellow (Φ) suggest that the leaf- purpling effect is independent of the state of phytochrome.

In 2013, Dean Kopsell and Carl Sams reported that a low dosage of blue LED light given near the end of a production cycle as sole-source lighting enhanced pigment and phytonutrient contents of broccoli microgreens (35). Michael Dzakovich in the Mitchell lab found no enhancement of tomato fruit-quality attributes with LED supplemental lighting in the greenhouse (36). Because LED supplementation occurred within a background of solar light, other wavelengths, such as green, may have negated any promotive effects that blue or red otherwise may have to enhance tomato-quality attributes.

There also is considerable interest in controlling plant growth with light from white LEDs, which actually are blue LEDs with a phosphor inside the lens that generates secondary wavelengths, making white LEDs broad- band rather than narrow-band. Bruce Bugbee's lab has examined cool-white, neutral-white, and warm-white LEDs (37). For both seedlings as well as more mature plants, the cooler the white (more blue content), the smaller the plants, both from a stem-elongation as well as a leaf - expansion perspective (Fig. 22). Will white LEDs substitute for red, blue, and far- red? They may for some applications, but not for others. It is good to have options!

The Bugbee lab also has compared the economics of different light sources, including HPS as well as LEDs. (38). The best of each light source have equivalent photon efficiencies, expressed as micromoles of photon output per unit of electrical energy input (Fig. 23). Among the LEDs, the best blues and reds have the highest photon efficiency followed by the whites, then it drops down for the far-reds and the greens, and UV-A-emitting LEDs are still very inefficient. As the drive current increases, the efficiency of all LEDs decreases. This could be partially related to operating temperature, so it is critically important to operate LEDs at nominal drive current to preserve performance and lifespan, and also to heat sink as needed.

Early in the evolution of LED technology, a pattern emerged predicting that for every decade, the efficiency of photon output from LEDs would increase 30-fold while the cost of production would go down 10-fold (Fig. 24, Ref. 39). It is not clear if these so-called "Haitz Law" (after the author) trends are still extrapolating linearly today, but LED capabilities have indeed improved substantially in recent times. Blue LEDs were just 8-11% efficient converting electricity to photons about 6 years ago. Today they are 50% efficient. Red LEDs were on a plateau of about 30% efficiency for a long time; now the best reds also are 50% efficient or better. It has been speculated that the theoretical limit for LED efficiency is about 80%. Although manufacturing costs are going down, the capital investment for growers still is 3-5 times higher for LED arrays than for equivalent-coverage HPS fixtures. This is probably the main issue holding back commercial growers from investing substantially in and adopting LED technology. Of course, there are no hazardous material issues associated with disposal, and light distribution out of LED arrays can be controlled by lens angle.

Significant milestones in the history of LEDs (prepared by Bob Morrow) since their invention in 1962 to the first publications on LEDs for plant growth in 1990 to award of the Nobel prize in 2014 for development of the blue LED in the 1990s cover more than half a century of total history and somewhat more than half of that period for plant applications per se (Fig. 25). This proceedings is neither exhaustive nor all-inclusive. It reflects both the space limitations of this proceedings as well as the impressions and interests of the author.

Literature Cited

1. http://en.wikipedia.org/wiki/Light-emitting_diode
2. http://hyperphysics.phy-astr.gsu.edu/hbase/electronic/led.html
3. Bula R.J., R.C. Morrow, T.W. Tibbitts, and D.J. Barta, R.W. Ignatius, and T.S. Martin. 1991. LEDs as a radiation source for plants. HortScience 26: 203-205.
4. Hoenecke, M.E., R.J. Bula, and T.W. Tibbitts. 1992. Importance of 'blue' photon levels for lettuce seedlings grown under red-light-emitting diodes. HortScience 27(5): 427-430.
5. McCree, K.J. 1972. The action spectrum, absorptance, and quantum yield of photosynthesis in crop plants. Agr. Meteorol. 9: 191-216.
6. Sager, J.C., W.O. Smith, J.L. Edwards, and K.L. Cyr. 1988. Use of spectral data to determine photosynthetic efficiency and phytochrome photoequilibria. Trans. ASAE 31(5): 1882-1889.

7. Miyashita, Y., Y. Kitaya, T. Kozai, and T. Kimura. 1995. Effects of red and far-red light on the growth and morphology of potato plantlets in vitro: using light emitting diode as a light source for micropropagation. Acta Hort. 393:189- 194.

8. Tennessen, D.J., R.L. Singsaas, and T.D. Sharkey. 1994. Light-emitting diodes as a light source for photosynthesis research. Photosynth. Res. 39:85–92.

9. Brown, C.S., A.C. Schuerger, and J.C. Sager. 1995. Growth and photomorphogenesis of pepper plants grown under red light-emitting diodes supplemented with blue or far-red illumination. J. Amer. Soc. Hort. Sci. 120:808–813.

10. Yanagi, T., K. Okamoto, and S. Takita. 1996. Effect of blue and red light intensity on photosynthetic rate of strawberry leaves. Acta Hort. 440:371–376.

11. Matsuda, R., K. Ohashi-Kaneko, K. Fujiwara, E. Goto, and K. Kurata. 2004. Photosynthetic characteristics of rice leaves grown under red light with or without supplemental blue light. Plant & Cell Physiol. 45:1870–1874.

12. Barta et al. 1992. Evaluation of LED characteristics for space-based irradiation source. Adv. Space Res. 12: 141-149.

13. Morrow et al. 1995. Plant response in the ASTROCULTURE Plant flight experiment unit. SAE Technical Paper 951624.

14. Zhou, W. 2005. Advanced Astroculture! plant growth unit: Capabilities and performances. SAE Technical Paper 2005-01-2840.

15. Croxdale et al. 1997. Structure of potato tubers formed during spaceflight. J. Expt. Bot. 48: 2037-2043

16. Stankovic, B., W. Zhou, and B.M. Link. 2002. Seed to seed growth of Arabidopsis thaliana on the International Space Station. SAE Technical Paper 2002-01-2284.

17. Tripathy, B.C. and C.S. Brown. 1995. Root-shoot interaction in the greening of wheat seedlings grown under red light. Plant Physiol. 107:407–411.

18. Goins, G.D., N.C. Yorio, M.M. Sanwo, and C.S. Brown. 1997. Photomorphogenesis, photosynthesis, and seed yield of wheat plants grown under red light-emitting diodes (LEDs) with and without supplemental blue lighting. J. Expt. Bot. 48:1407–1413.

19. Yorio, N.C., G.D. Goins, H.R. Kagie, R.M. Wheeler, and J.C. Sager. 2001. Improving spinach, radish, and lettuce growth under red light-emitting diodes (LEDs) with blue light supplementation. HortScience 36:380–383.

20. Morrow, R.C. and T.W. Tibbitts. 1988. Evidence for involvement of phytochrome in tumor development in plants. Plant Physiol. 88: 1110 - 1114.

21. Morrow, R.C. and R.M. Wheeler. 1997. Physiological Disorders. P. 133-141, In: R.W. Langhans and T.W. Tibbitts (eds), A growth chamber manual, 2nd ed. Iowa State University Press. North Central Regional Research Publication 340.

22. Kim, H.H., G.D. Goins, R.M. Wheeler, and J.C. Sager. 2004b. Green-light supplementation for enhanced lettuce growth under red- and blue light- emitting diodes. HortScience 39:1617–1622.

23. Folta, K.M. and S.A. Maruhnich. 2007. Green light: A signal to slow down or stop. J. Expt. Bot. 58:3099–3111.
24. Massa, G.D., H.-H. Kim, R. M. Wheeler, and C.A. Mitchell. 2008. Plant productivity in response to LED lighting. HortScience 43 (7): 1951-1956.
25. Frantz, J., C. Chun, R. Joly, and C. Mitchell. 1998. Intracanopy lighting of cowpea canopies in controlled environments. Life Support Biosphere Sci. 5: 183-189.
26. Frantz, J., R. Joly, and C. Mitchell. 2000. Intracanopy lighting influences radiation capture, productivity, and leaf senescence in cowpea canopies. J. Amer. Soc. Hort. Sci. 125:694-701.
27. Frantz, J., R. Joly, and C. Mitchell. 2001. Intracanopy lighting reduces electrical energy utilization by closed cowpea stands. Life Support & Biosphere Sci. 7: 283-290.
28. Poulet, L., G.D. Massa, R.C. Morrow, C.M. Bourget, R. M. Wheeler, and C.A. Mitchell. 2014. Significant reduction in energy for plant-growth lighting in space using targeted LED lighting and spectral manipulation. Life Sciences in Space Research 2: 43-53.
29. Mitchell, C., J. Burr, M. Dzakovich, C. Gomez, R. Lopez, R. Hernandez, C. Kubota, C. Currey, Q. Meng, E. Runkle, C. Bourget, R. Morrow, and A.J. Both. 2015. LEDs in Horticulture. Horticultural Reviews 43: 1-87.
30. Gomez, C., R.C. Morrow, C.M. Bourget, G.D. Massa, and C.A. Mitchell. 2013. Comparison of intracanopy light-emitting diode towers and overheard high- pressure sodium lamps for supplemental lighting of greenhouse-grown tomatoes. HortTechnology 32(1): 93-98.
31. Gomez, C. and C.A. Mitchell. 2014. Supplemental lighting for greenhouse-grown tomatoes: intracanopy LED towers vs. overhead HPS lamps. Acta Hort. 1037: 855-862.
32. Gomez, C. and C.A. Mitchell. 2015. Growth responses of tomato seedlings to different spectra of supplemental lighting. HortScience 50 (1): 112-118.
33. Gomez, C. and C.A. Mitchell. 2016. Physiological and productivity responses of high-wire tomato as affected by supplemental light source and distribution within the canopy. J. Amer. Soc. Hort. Sci. 141(2): 196-208.
34. Stutte, G.W., S. Edney, and T. Skerrit. 2009. Photoregulation of bioprotectant content of red leaf lettuce with light-emitting diodes. HortScience 44: 79-82.
35. Kopsell, D.A. and C.E. Sams. 2013. Increases in shoot tissue pigments, glucosinolates, and mineral elements in sprouting broccoli after exposure to short-duration blue light from light-emitting diodes. J. Amer. Soc. Hort. Sci. 138 (1): 31-37.

36. Dzakovich, M.P., C. Gomez, and C.A. Mitchell. 2015. Tomatoes grown with light- emitting diodes or high-pressure sodium lights have similar fruit-quality attributes. HortScience 50 (10): 1498-1502.

37. Cope, K.R. and B. Bugbee. 2013. Spectral effects of 3 types of white LEDs on plant-growth and development: absolute vs. relative amounts of blue light. 2013. HortScience 48(4): 504-509.

38. Nelson, J.A. and B. Bugbee, 2014. Economic analysis of greenhouse lighting: light- emitting diodes vs. high-pressure sodium fixtures. PloS ONE 9(6): e991doi:1010 137/ journal pone. 0099010

39. Drennen, T., R. Haitz, and J. Tsao. 2001. A market diffusion and energy impact model for solid-state lighting. Sandia National Laboratories, SAND2001–2830J. August

Acknowledgements
Thanks to Bob Morrow, Orbital Technologies Corporation (A subsidiary of Sierrra Nevada; Ray Wheeler, NASA Kennedy Space Center; and to Chieri Kubota, Controlled Environment Agriculture Center, University of Arizona; all for images and information that greatly assisted in assembling this historical perspective of LEDs and plant growth.

Fig. 1

What is an LED?

- Light Emitting Diode (first practical red LED invented in 1962)
- No filament, illuminated solely by movement of electrons in a semiconductor material
- Electrons cross junction recombine with electron holes, release energy as photons (electroluminescence)
- Color determined by the energy gap of semiconductor (based on semiconductor chemical composition)

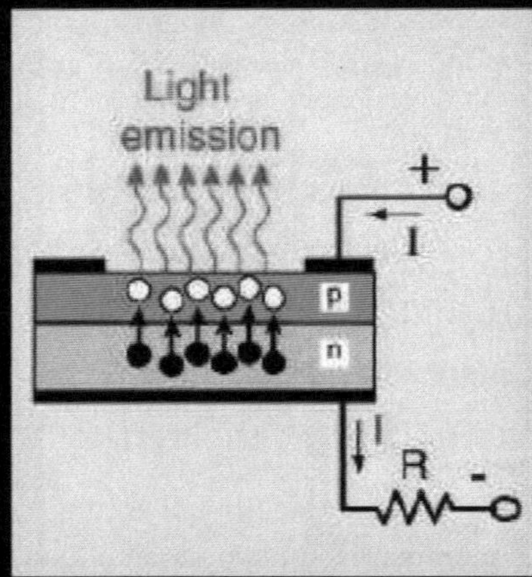

http://en.wikipedia.org/wiki/Light-emitting_diode
http://hyperphysics.phy-astr.gsu.edu/hbase/electronic/led.html

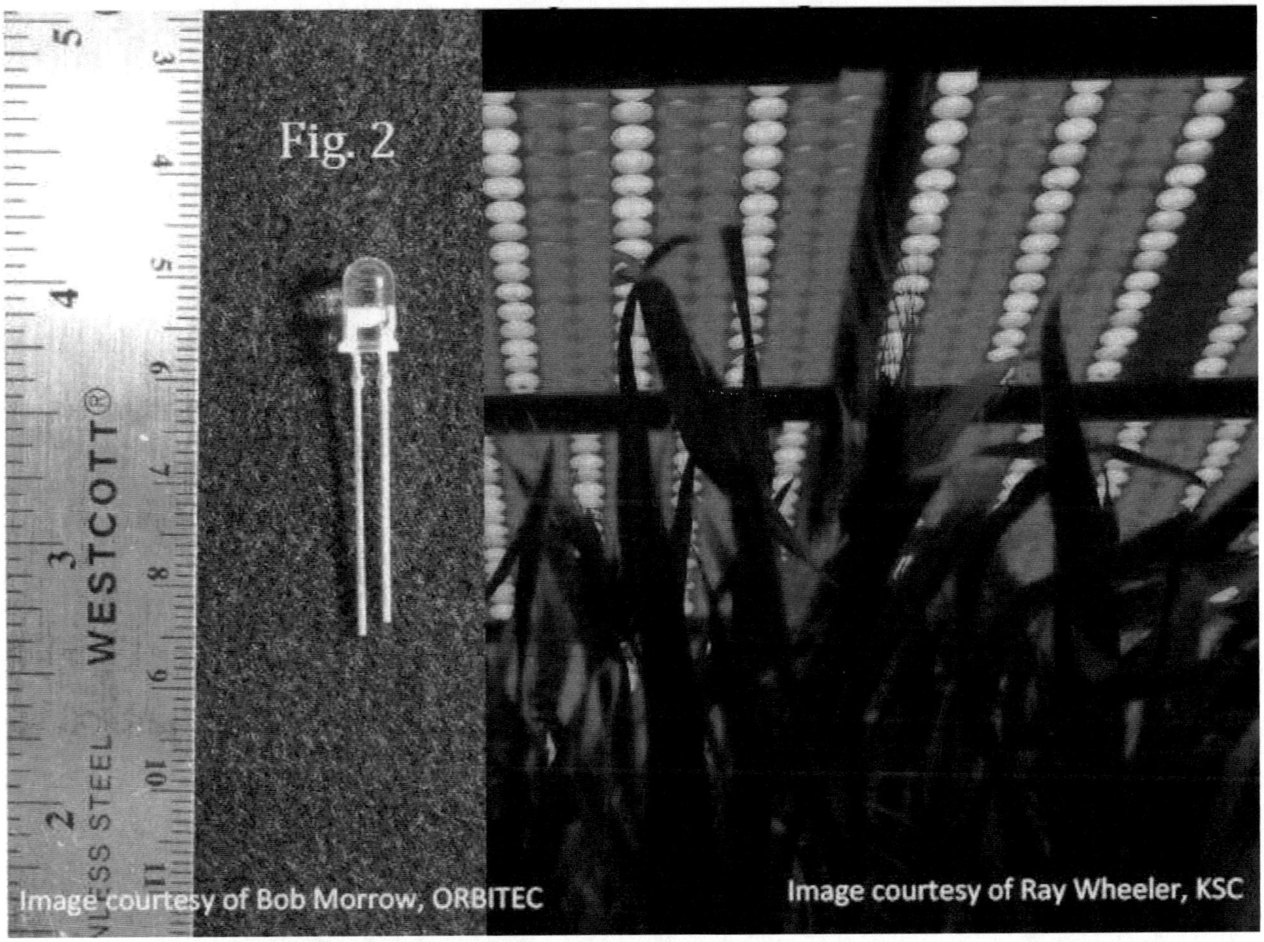
Fig. 2
WESTCOTT®
Image courtesy of Bob Morrow, ORBITEC
Image courtesy of Ray Wheeler, KSC

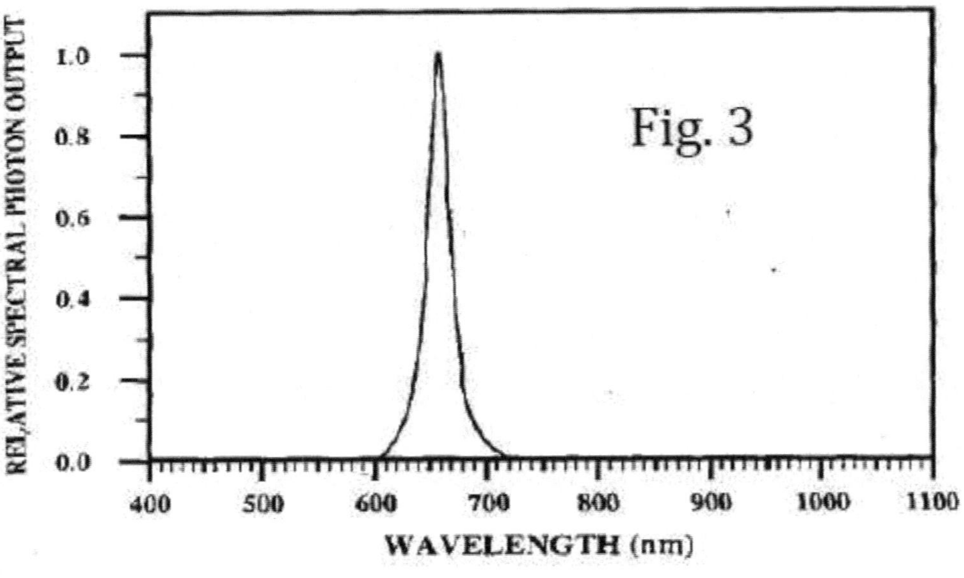
Fig. 3
RELATIVE SPECTRAL PHOTON OUTPUT
1.0
0.8
0.6
0.4
0.2
0.0
400
500
600
700
800
900
1000
1100
WAVELENGTH (nm)

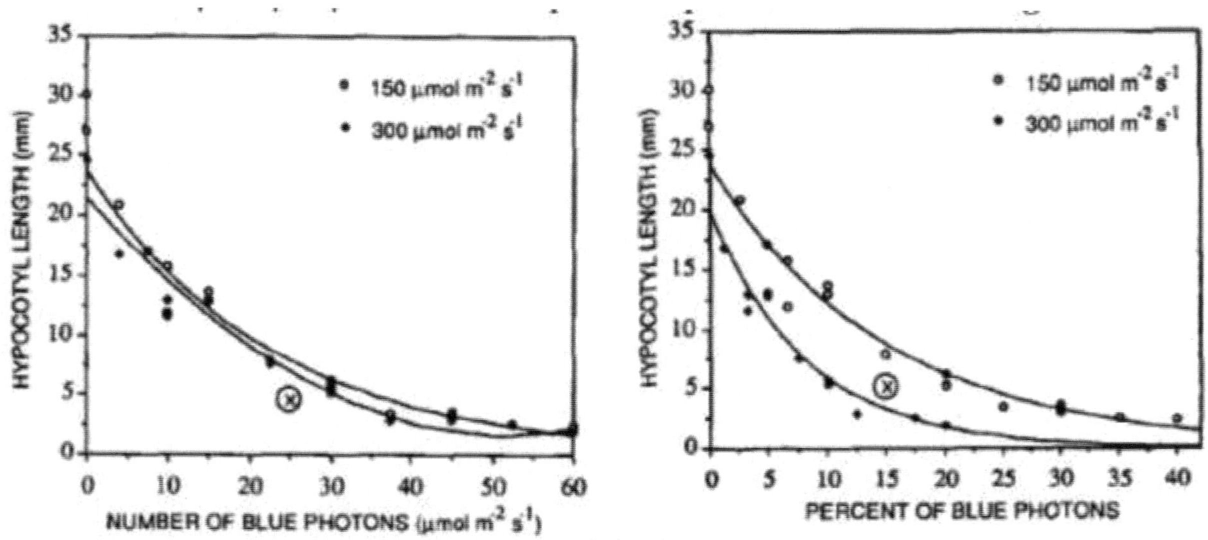
150 μmol m⁻² s⁻¹
300 μmol m⁻² s⁻¹
HYPOCOTYL LENGTH (mm)
NUMBER OF BLUE PHOTONS (μmol m⁻² s⁻¹)
150 μmol m⁻² s⁻¹
300 μmol m⁻² s⁻¹
HYPOCOTYL LENGTH (mm)
PERCENT OF BLUE PHOTONS

Fig. 4

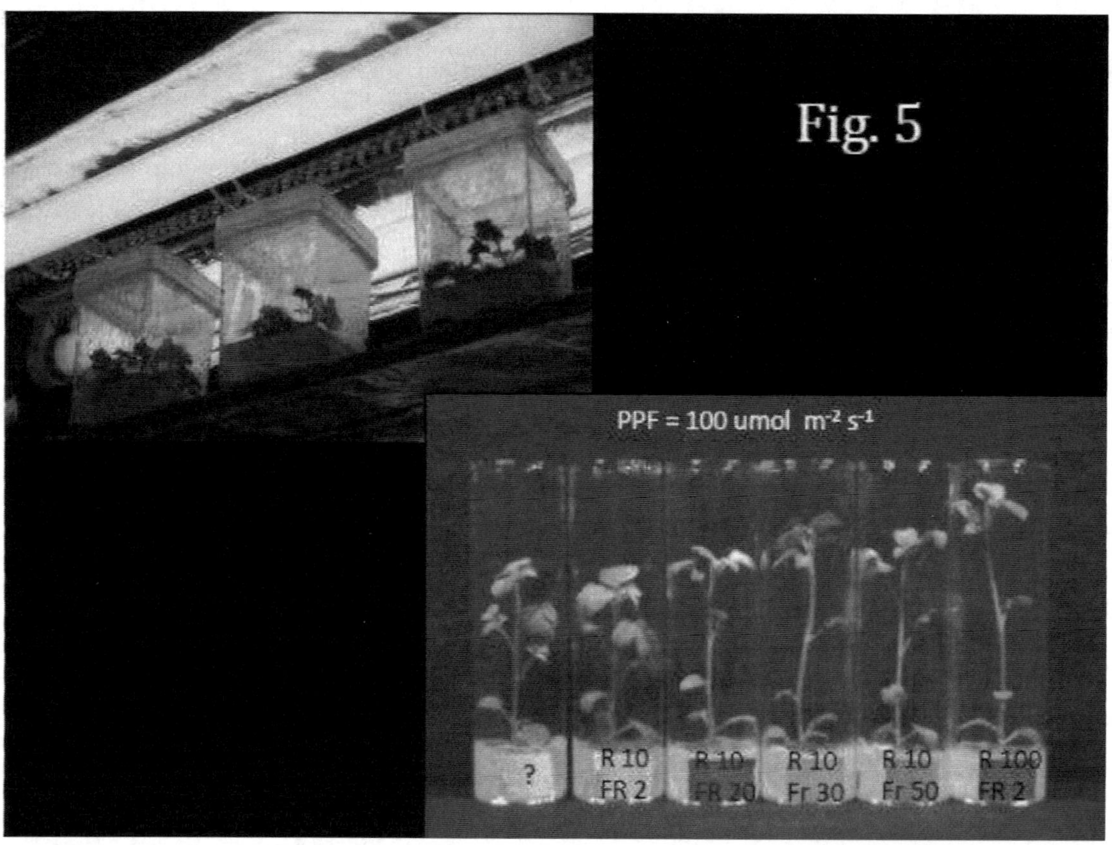
PPF = 100 umol m-2 s-1
?
R 10
FR 2
R 10
FR 20
R 10
Fr 30
R 10
Fr 50
R 100
FR 2

Fig. 5

Plant Species grown with LEDs in preparation for spaceflight
'Grand Rapids' lettuce
Dwarf rapid cycling brassica
Grown for 21 days under 16:8 photoperiod at 295 µmol/m²/s red + 30 µmol/m²/s blue fl

Fig. 7

R+ B LED lighting
in a growth chamber
Cowpea leaf
Massa and Mitchell
Low-level solar in a winter greenhouse
Tomato leaf
Gomez and Mitchell

Photobleaching from too-close LED placement
Fig. 8

Effects of green light on plant growth

Photo courtesy of Ray Wheeler, KSC

Fig. 9

Table 2. Influence of light quality on leaf area, specific leaf area (SLA), shoot fresh weight (shoot FW), shoot dry weight (shoot DW), single leaf net photosynthesis (Pn), total chlorophyll (Chl) content, and canopy leaf temperature at 28 d after planting.

Parameter	Treatment[z]			
	RB	RGB	GF	CWF
Leaf area (cm^2)	524.8 b[y]	689.9 a	419.2 c	595.3 b
SLA ($m^2 \cdot kg^{-1}$)	34.1 b	30.1 c	49.4 a	33.9 b
Shoot FW (g)	24.6 b	35.7 a	15.3 c	26.9 b
Shoot DW (g)	1.54 b	2.26 a	0.83 c	1.76 b
Pn ($\mu mol\ CO_2/m^2/s$)	9.3 a	8.6 a	7.4 b	8.8 a
Chl ($g \cdot m^{-2}$)	0.21 a	0.21 a	0.18 a	0.20 a
Canopy leaf temperature (°C)	20.5 a	20.4 a	20.7 a	20.3 a

[z]See Fig. 1 and Table 1 for spectral characteristics.

[y]Mean comparison within a row by Duncan's multiple range test, $P = 0.05$. Means with same letter are not significantly different.

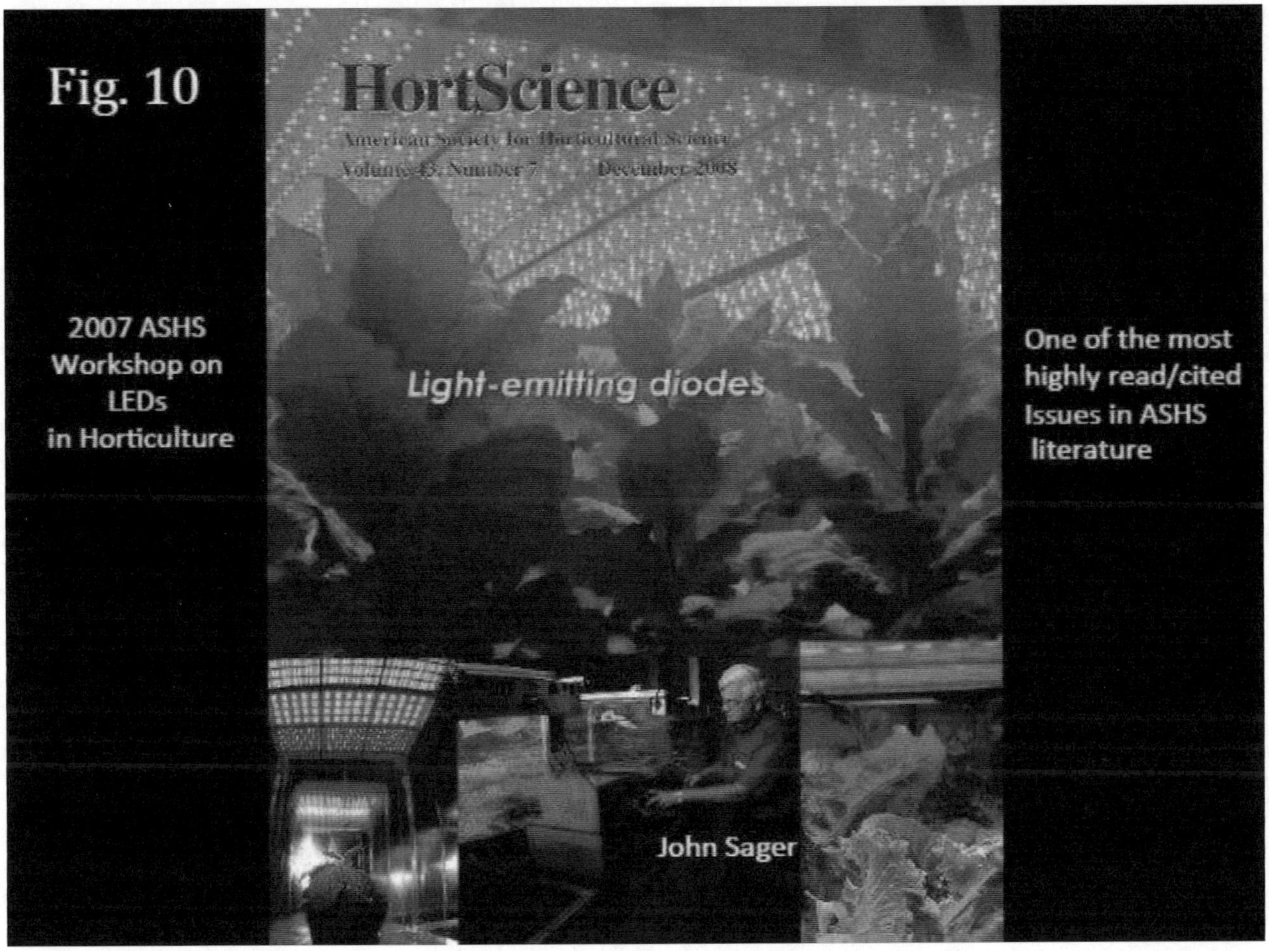
Fig. 10
2007 ASHS Workshop on LEDs in Horticulture
HortScience
Light-emitting diodes
John Sager
One of the most highly read/cited Issues in ASHS literature

"Veggie" Plant-Growth Unit for International Space Station (ISS) – launched on SpaceX-4, September 2014
Expandable accordion walls
Deployed on ISS
Figure 5.10. Astronaut Steve Swanson preparing to harvest lettuce grown under LEDs in the VEGGIE plant-growth chamber on the International Space Station (Image credit: NASA).
Image courtesy of Bob Morrow
Fig. 11

Fig. 12 C.A. Mitchell: LED "Lightsicle" Concept, 1996

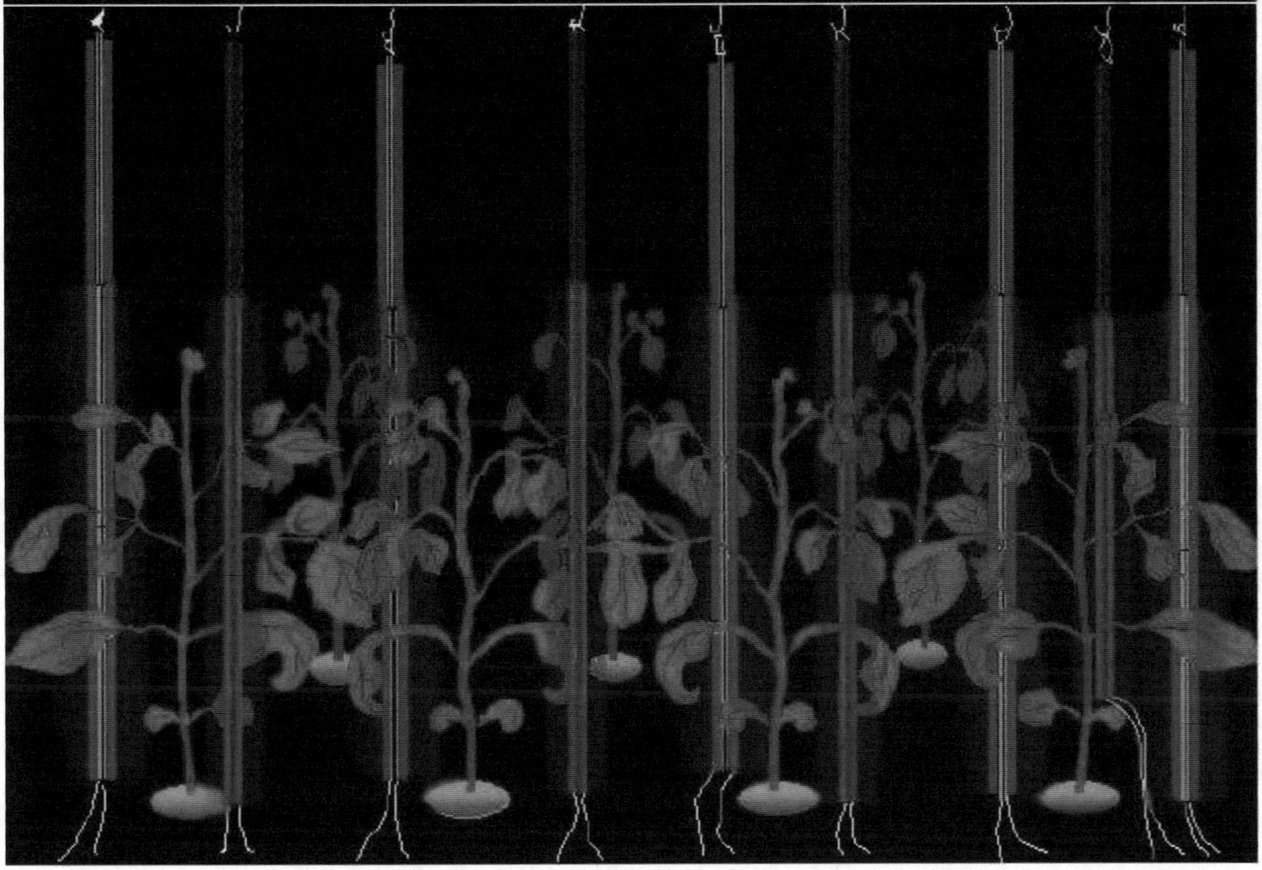

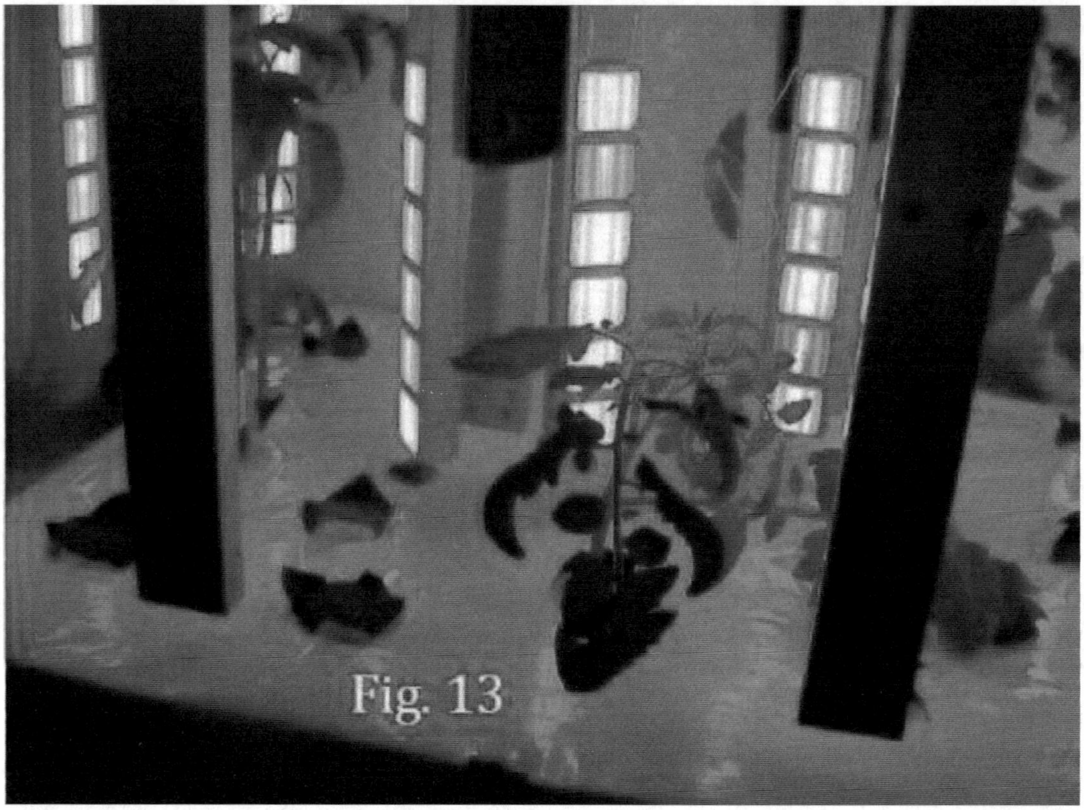
Fig. 13

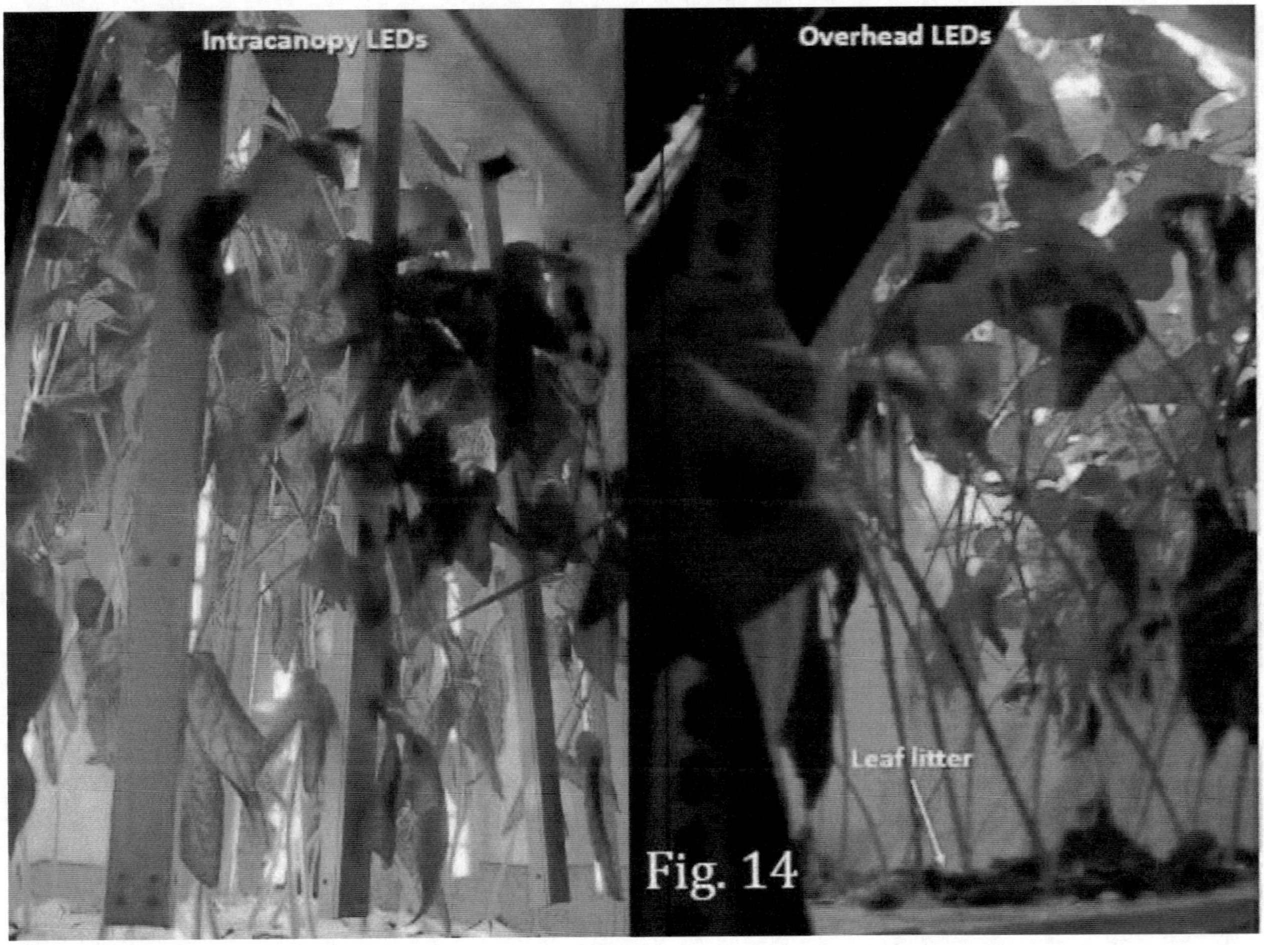
Intracanopy LEDs
Overhead LEDs
Leaf litter
Fig. 14

Fig. 15

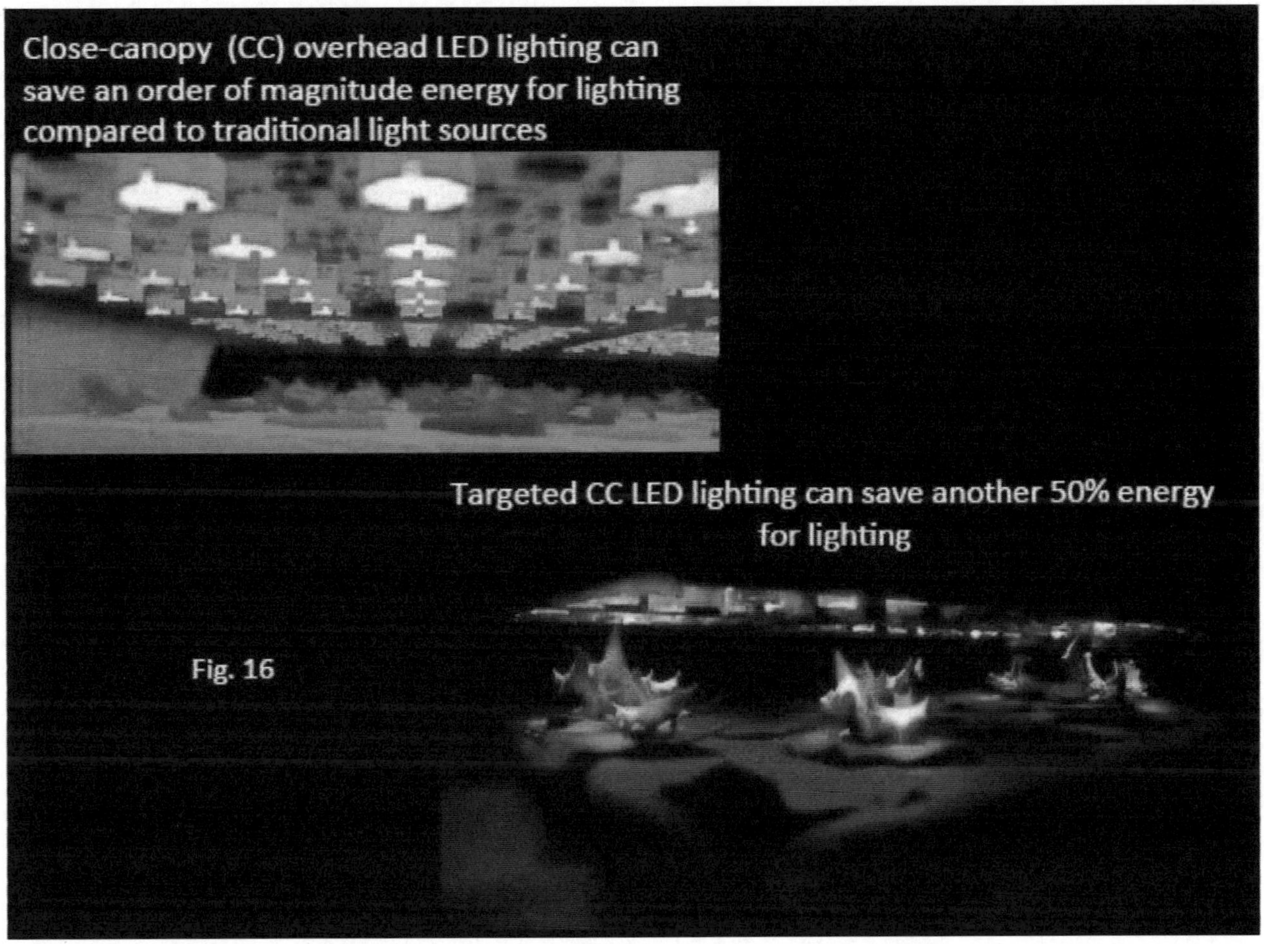
Close-canopy (CC) overhead LED lighting can save an order of magnitude energy for lighting compared to traditional light sources
Targeted CC LED lighting can save another 50% energy for lighting

Fig. 16

Fig. 17

Light-Emitting Diodes in Horticulture

Cary A. Mitchell, Michael P. Dzakovich, Celina Gomez, and Roberto Lopez
Department of Horticulture & Landscape Architecture
Purdue University
West Lafayette, IN USA

John F. Burr
Krannert School of Business Management
Purdue University
West Lafayette, IN USA

Richardo Hernández and Chieri Kubota
School of Plant Sciences
The University of Arizona
Tucson, AZ USA

Christopher J. Currey
Department of Horticulture
Iowa State University
Ames, IA USA

Qingwu Meng and Erik S. Runkle
Department of Horticulture
Michigan State University
East Lansing, MI USA

Christopher M. Bourget and Robert C. Morrow
Orbital Technologies Corporation
Madison, WI USA

Arend J. Both
Department of Environmental Sciences
Rutgers University
New Brunswick, NJ USA

Horticultural Reviews, Volume 43, First Edition. Edited by Jules Janick.

Fig. 18 LED Intracanopy Lighting
(from the work of Celina Gomez and Cary Mitchell, Purdue University)
Intracanopy lighting tower technology developed by ORBITEC

Tomato fruit yield with or without supplemental lighting and energy consumption for IC LED lighting vs. OH HPS lighting

Fig. 19

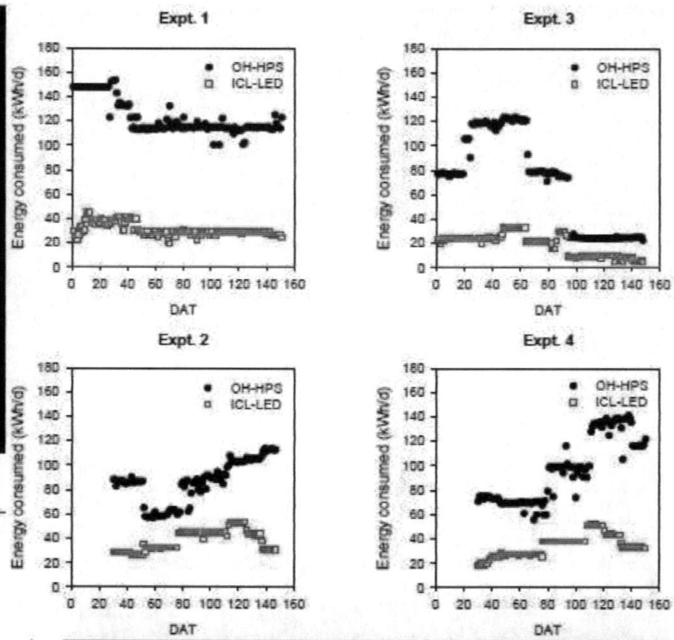

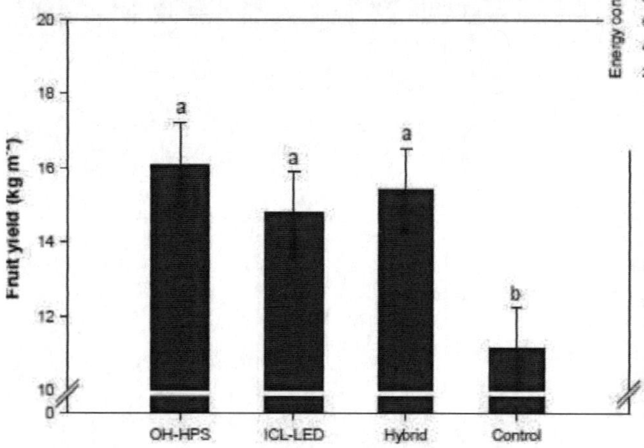

Expts. 1 & 3 = winter to summer
Expts. 2 & 4 = summer to winter

Fig. 20

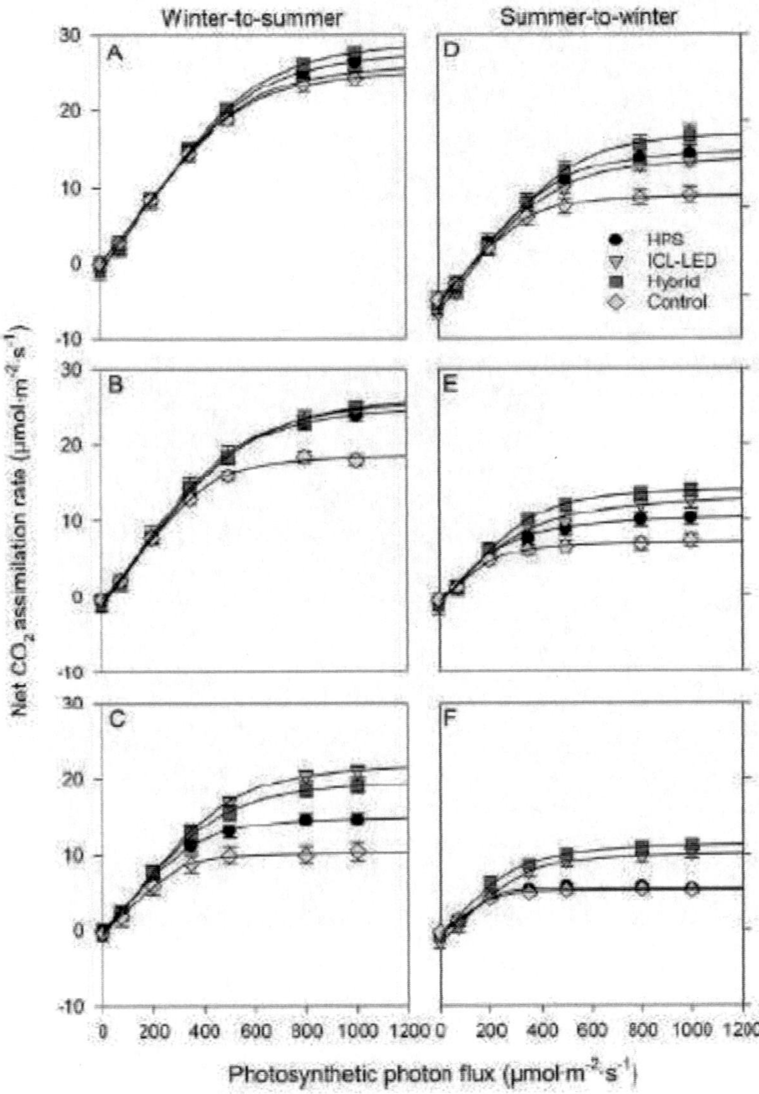
Winter-to-summer
Summer-to-winter
A
B
C
D
E
F
HPS
ICL-LED
Hybrid
Control
Net CO2 assimilation rate (µmol·m-2·s-1)
Photosynthetic photon flux (µmol·m-2·s-1)
30
20
10
0
-10
0
200
400
600
800
1000
1200

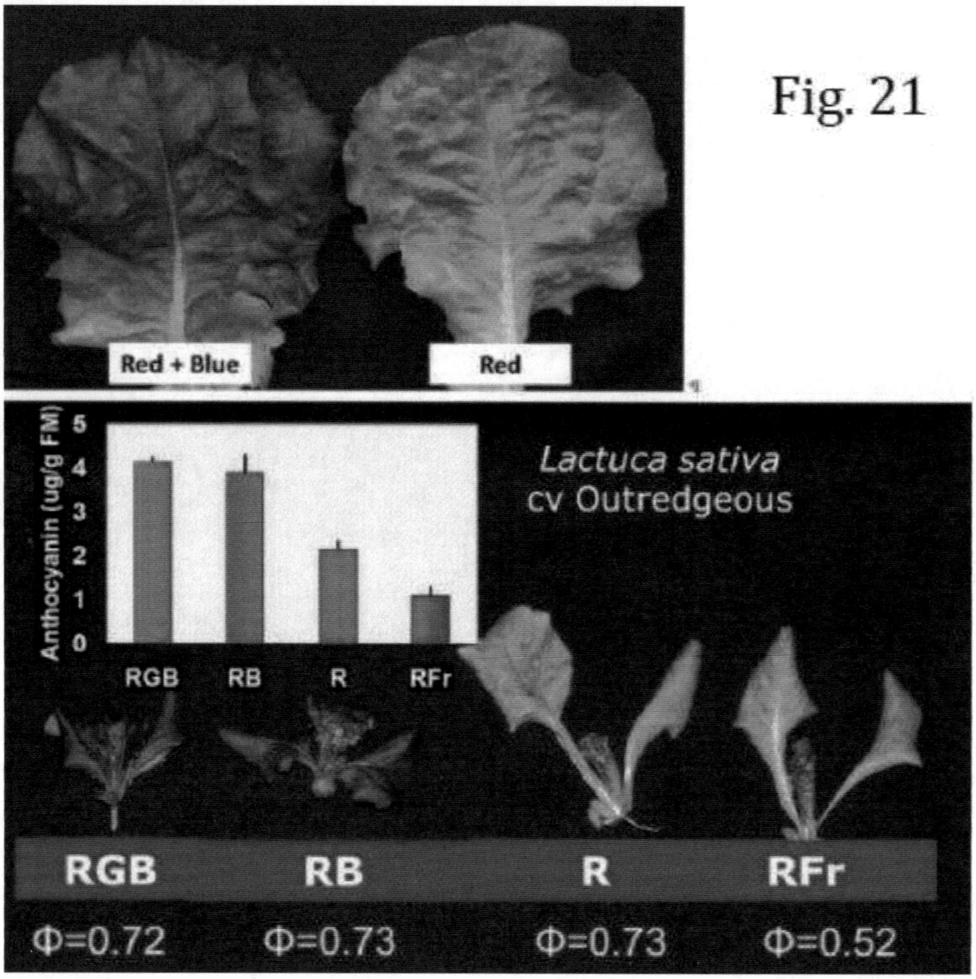
Red + Blue
Red
Anthocyanin (ug/g FM)
0
1
2
3
4
5
RGB
RB
R
RFr
Lactuca sativa
cv Outredgeous
RGB
RB
R
RFr
Φ=0.72
Φ=0.73
Φ=0.73
Φ=0.52

Fig. 21

A role for White LEDs in plant-growth regulation

Fig. 22

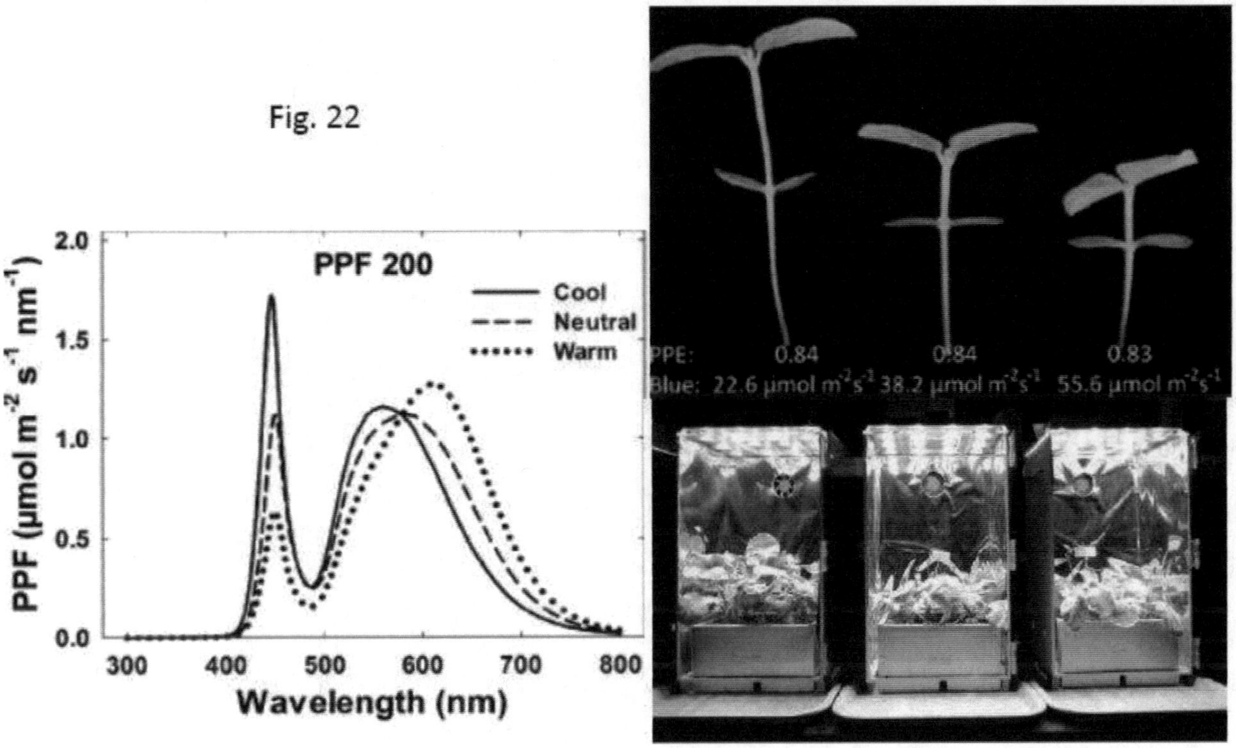

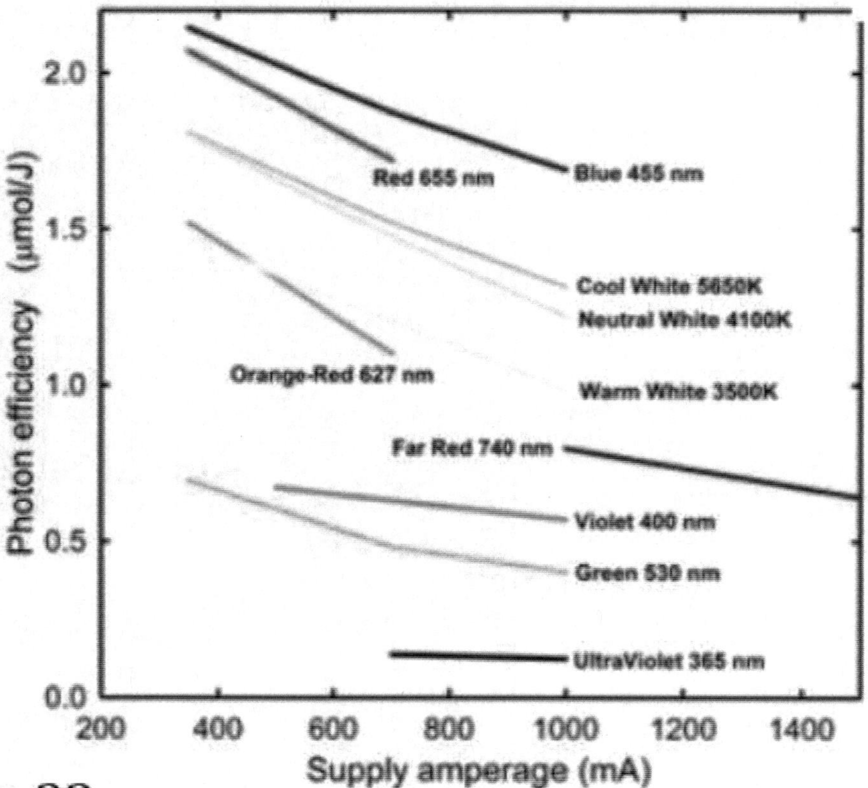

Photon efficiency (µmol/J)
2.0
1.5
1.0
0.5
0.0
200
400
600
800
1000
1200
1400
Supply amperage (mA)
Red 655 nm
Blue 455 nm
Cool White 5650K
Neutral White 4100K
Orange-Red 627 nm
Warm White 3500K
Far Red 740 nm
Violet 400 nm
Green 530 nm
UltraViolet 365 nm

Fig. 23

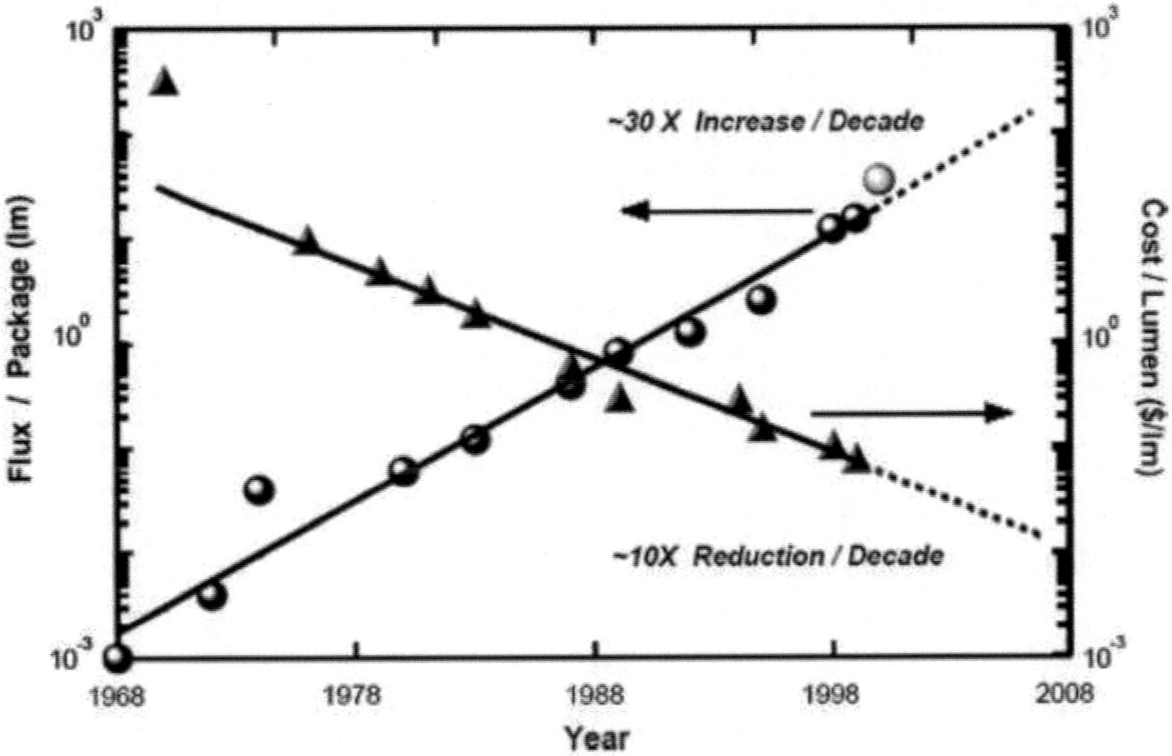

Fig. 3. A comparison of light-emitting diode costs and light output versus time.

Fig. 24

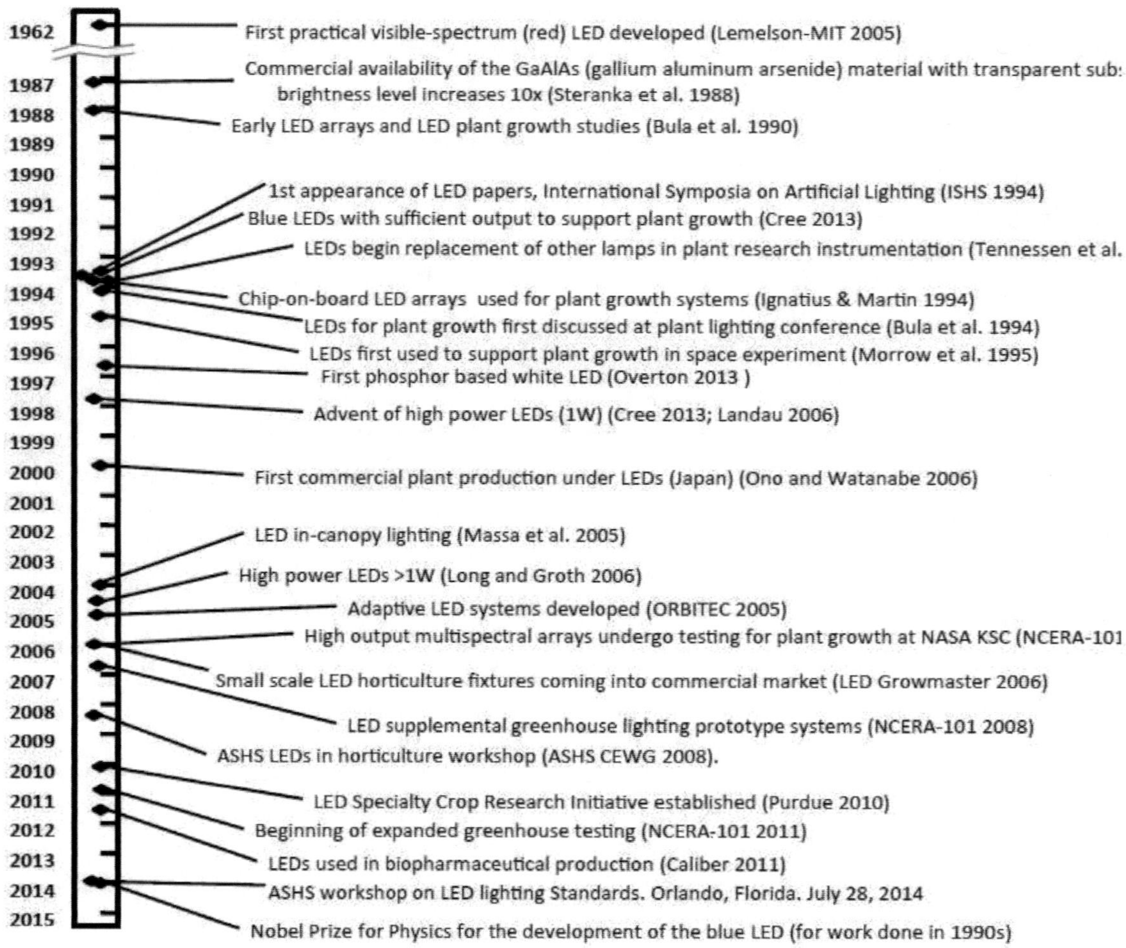
1962
1987
1988
1989
1990
1991
1992
1993
1994
1995
1996
1997
1998
1999
2000
2001
2002
2003
2004
2005
2006
2007
2008
2009
2010
2011
2012
2013
2014
2015
First practical visible-spectrum (red) LED developed (Lemelson-MIT 2005)
Commercial availability of the GaAlAs (gallium aluminum arsenide) material with transparent sub:
brightness level increases 10x (Steranka et al. 1988)
Early LED arrays and LED plant growth studies (Bula et al. 1990)
1st appearance of LED papers, International Symposia on Artificial Lighting (ISHS 1994)
Blue LEDs with sufficient output to support plant growth (Cree 2013)
LEDs begin replacement of other lamps in plant research instrumentation (Tennessen et al.
Chip-on-board LED arrays used for plant growth systems (Ignatius & Martin 1994)
LEDs for plant growth first discussed at plant lighting conference (Bula et al. 1994)
LEDs first used to support plant growth in space experiment (Morrow et al. 1995)
First phosphor based white LED (Overton 2013)
Advent of high power LEDs (1W) (Cree 2013; Landau 2006)
First commercial plant production under LEDs (Japan) (Ono and Watanabe 2006)
LED in-canopy lighting (Massa et al. 2005)
High power LEDs >1W (Long and Groth 2006)
Adaptive LED systems developed (ORBITEC 2005)
High output multispectral arrays undergo testing for plant growth at NASA KSC (NCERA-101
Small scale LED horticulture fixtures coming into commercial market (LED Growmaster 2006)
LED supplemental greenhouse lighting prototype systems (NCERA-101 2008)
ASHS LEDs in horticulture workshop (ASHS CEWG 2008).
LED Specialty Crop Research Initiative established (Purdue 2010)
Beginning of expanded greenhouse testing (NCERA-101 2011)
LEDs used in biopharmaceutical production (Caliber 2011)
ASHS workshop on LED lighting Standards. Orlando, Florida. July 28, 2014
Nobel Prize for Physics for the development of the blue LED (for work done in 1990s)

Fig. 25

SPECTRAL CONTROL: IMPACTS OF NARROW AND BROAD SPECTRA LED AND LEGACY LIGHT IRRADIATION ON PLANT MORPHOLOGY AND YIELD

J. Mastin and B. Koontz
Venntis Technologies, 1261 S Waverly Rd, Holland, MI 49423 USA

This is our first time having the pleasure of joining you all here, but I think it's safe to assume that the emphasis on lighting in the PGR world has taken a substantially more prominent place recently. This new world of especially LED lighting to control plant morphology is an exciting one and has clearly only begun to be unveiled.

At Venntis Technologies and TotalGrow Lights, our emphasis has been on a more difficult to quantify aspect of LED-based grow lighting – broad spectrum lights. LEDs have been convenient horticultural lighting testing tools for decades now because they can apply narrow regions of the electromagnetic spectrum with a high level of control. As a result, there is a relative wealth of information about the impacts of different red to blue ratios, possibly with a green or red far red element included. Unfortunately, the difficulty in quantifying a spectrum and the amount of confounding variables increases exponentially when you try to address broad spectra instead of just narrow spectra sources.

As early as the 1950s, Robert Emerson was producing data, now expounded upon by researchers like keynote speaker Kevin Folta, showing that the photosynthetic and photomorphological responses of a plant are often not the sum of the spectral components. Rather, spectral regions and their activated photoreceptors interact in ways that can be extremely important towards the end efficiency and impacts of a light spectrum. We continue to learn about the interactions between different regions, such as blues and greens, blues and far reds, and reds and far reds. While less understood, ignoring these interactions misses the potential to balance out the positive and negative impacts of each spectral region while capitalizing on enhancement effects. The questions abound on how to best balance leaf expansion to capture more light with using the most absorbable wavelengths and those that best trigger nutritional benefits. The accurate, but not so helpful answer, is that it all depends --- on crop, cultivar, specific goals, temperature, nutrition, and several other variables. Despite the convoluted nature of this topic, there are some clear trends we are seeing.

In general increasing blue content and increasing the red to far red ratio has an effect along the lines of a chemical PGR with the added benefits of increased phytonutrient and secondary metabolite concentrations and enhanced pigmentation. Many studies on cryptochrome and phytochrome responses have demonstrated these general trends.

However, as the Bruce Bugbee study here alludes, there are potential risks when overemphasizing the PGR-like effects of narrow waveband LEDs. The red and especially blue photons that trigger the growth regulating effects are efficiently absorbed and used for photosynthesis. However, the resultant reduction in leaf area often correlates to reduced overall growth when there is incomplete canopy coverage due to less radiation capture. When growers and researchers include more broad spectral lighting sources, positive overall growth results often follow at least in part due to the increase leaf expansion triggered at least in part due to green light. Here are just a few studies showing improved overall growth when plants received a more complete spectrum than the simple red-blue mixes often used.

As mentioned, our own research and experience has naturally included broad spectra. In the interest of time, I'll go in depth on just one of the more interesting sets of tests. Dr. Youyoun Moon out of West Virginia University has been conducting research on the full gamut of lighting types for about 2 years now. He has included tomatoes, kale, lettuce, basil, and petunias under several lighting types, including a variety of typical LED spectra, induction lights, high pressure sodium lights, metal halides, light emitting plasma and our TotalGrow lights.

Preliminary testing was on basil, petunias and tomatoes under HPS, metal halide, induction, 3 narrower band LED spectra and TotalGrow's broad spectrum. Minimal data was collected, but by observations, Dr. Moon reported that, "All plants we have tested (petunia, basil, and tomato) looked healthier under your TotalGrow™ lighting and used far less energy than the competing light sources." In addition, the broader spectral light sources were far more pleasant to work around as plants and their surrounding look much more natural.

The more in depth studies involved lettuce under HPS, metal halide, 3 narrower LED spectra and the broad TotalGrow spectrum, and kale in a similar setup with one more LED type. Testing was conducted at 200 and/or 400 μmol/m2/s light intensities. To remove the confounding effect of fertilization, plants were only grown until soil nutrients became limiting rather than adding a fertilization regimen.

The results of this test are a great example of the need for a grower to decide on their priorities. Before diving into any numbers, here is a practical question for growers to consider when trying to wade through the piles of what can often appear to be conflicting information. What are your goals and priorities? What will help win and maintain the customers you want and generate the revenue you need from them?

- Yield
- Nutritional content
- Coloration (which is often correlated to nutritional content
- Electrical Efficiency
- Form Factor – Remember the variations in luminaire styles even in this test from the previous slide
- Working Environment Aesthetics (broader, whiter spectra are generally helpful)
- Upfront Costs
  - o Lights
  - o Electrical Infrastructure
  - o Rebates
  - o HVAC – less efficient lights and those that radiate their heat require more cooling or less heating
- Other Factors?

Back to the lettuce results. The colors on these plots are very representative of the spectra of their respective light sources. Looking at the yields, the results suggest the greatest mass per plant resulting from the broad TotalGrow spectrum that still emphasizes the blue and red regions, followed by the more blue and green intensive broad metal halide spectrum and the narrow LED spectrum that included the greatest white portion as well. A heavy blue narrow spectrum narrow LED light produced the lowest yield per plant.

On the basis of power usage, which is dependent on how well the exact setup utilizes each lighting type, the four LED-based lights all show their advantage over HID lighting.

Unfortunately, phytochemical and secondary metabolite analysis is still underway. Nonetheless, the images of the lettuce provide a strong indication of what to expect. HPS and metal halides, with the majority of their output in the green region, produce the most stretched plants with the least pigmentation. The narrower LED lights produced plants that were most compact, lowest in mass, but highest in pigmentation and assumedly secondary metabolites as a whole. The TotalGrow spectrum achieved a higher pigmentation level than the HIDs due to higher blue and red concentrations, and greater masses than the narrower LEDs, but less pigmentation than the narrow LEDs with less compact growth. As was previously stated, the best light for a grower will depend on their needs and the priorities.

We don't have as much data from Dr. Moon's kale testing, but here is a quick look at images that demonstrate similar trends. As you can see, there is a lot of variation in crop size and quality depending on the light source. The crops are a green kale – Dwarf Blue Curled, and a red kale – Red Russian. Tests were run at both 200 and 400 μmol/m2/s with 16 hour day/8 hour nights at 23-25 oC. Plants were only irrigated with water and grown to a baby green size which provided insufficient mass for phenolics testing, but the coloration is a strong indication of this measure of quality. The images show quite clearly that broad spectrum light sources like metal halide and Totalgrow have the potential to drive the growth of greater biomass and leaf area, while narrow spectrum lights sources have the potential for the highest concentration of phenolics, secondary metabolites or pigments. A grower's individual goals therefore need to be carefully considered when selecting a lighting strategy. In particular, when little or no sunlight is available the choice of light spectrum can have drastic impact on the growth rates and quality of crops.

COMPARISON OF COOL WHITE FLUORESCENT TO VARIED LED'S SPECTRAL QUALITIES FOR HEALING OF GRAFTED TOMATO PLANTS IN HEALING CHAMBERS

Jimmy Byrtus[1], Cristian Collado[2], Melanie Yelton[1], and Ricardo Hernández[2]
[1]LumiGrow, 1480 64th Avenue, Suite 150, Emeryville, CA 94608
[2]Department of Horticultural Sciences, North Carolina State University, Raleigh, NC

Grafted tomato seedlings propagation techniques must be optimized to increase production efficiencies. Grafting has several steps that include growing, grafting, healing, and acclimating grafted plants. Healing is considered the most critical step. Healing is defined as the joining of cambium between rootstock and scion to form vascular tissue and is commonly done under modified greenhouse conditions or inside healing chambers. Previous work determined that light during the healing period influences morphology and quality of grafted tomatoes. Cool white fluorescent (CWF) at a photosynthetic photon flux of 100 µmol·m-2·s-1 is recommended for healing chambers. Light emitting diodes (LED) make possible optimizing the spectral requirements for healing. During healing, the plants were exposed to six different light treatments. Four LED spectra with varied blue (B), green (G) and red (R) percent photon flux (PF) of 4B:5G:89R, 30B:2G:67R, 50B:50R and 76B:24R at 75 µmol·m-2·s-1 PF. CWF was used as a control at 75 µmol·m-2·s-1 (CWF-75), and 100 µmol·m-2·s-1 (CW-100) PF. Total height, stem diameter, leaf area and leaf number were measured pre and post healing on each grafted plant. The %-increase from pre to post healing was calculated for plant parameters. Plants in 4B:5G:89R treatment had 31% greater height %-increase than in all other treatments. No differences were observed between plants in all treatments for %-increase of leaf area. CWF-100 had 21% greater dry mass than 30B:2G:67R. Plants in CWF 100 treatment had 31% greater plant compactness than in 4B:5G:89R and 30B:2G:67R. Plants in 50B:50R, 76B:24R and CWF 75 were not different than in CWF-100 for plant compactness. This preliminary experiment shows that tomato plant quality during healing is comparable under LED treatments (50B:50R, 76B:24R) at 75 µmol·m-2·s-1 to CWF at 100 µmol·m-2·s-1.

Materials and Methods

Trust seeds were planted at a rate of 1,000 ·m-² and placed into a germination chamber at 28°C for 48 hours. After germination the plants were moved to the greenhouse to be grown for grafting. The environmental conditions for the greenhouse for both repetitions are in Table 1. The plants were grown in the greenhouse for 15 to 17 days dependent on repetition. Prior to grafting plants were selected to be of similar size for each treatment. Upon Grafting data collected was stem diameter (mm) approximately 5 mm below the graft union, initial height (mm), leaf area (mm²), and number of leaves. At the time of grafting the plants were placed into healing boxes at a density of 500 plants·m-2 under different treatments. The treatments are as follows, LED spectra with varied blue (B), green (G) and red (R) percent photon flux (PF) of 4B:5G:89R, 30B:2G:67R, 50B:50R and 76B:24R at 75 µmol·m-2·s-1 PF. CWF was used as a control at 75 µmol·m-2·s-1 (CWF-75), and 100 µmol·m-2·s-1 (CW-100) PF. The temperature for each treatment within the healing boxes is in Table 2. The healing boxes were sealed to maintain humidity levels at or near 100% for three days and gradually lowered for three more days. The plants were taken out of the healing boxes after 6 days and data was collected. After healing the following data was collected, leaf area (mm^2), stem diameter (mm), final height (mm), leaf number, wet mass (g), and dry mass (g). Plant compactness was calculated using Dry Mass (g)/final height (mm). SAS JMP 12, (Cary, NC) was used for all statistical analysis. The data from the two repetitions were combined due to no interactions being found between repetitions for reported results.

Results

It is shown in Figures 1, 2, 3 that there was no difference for initial height, initial stem diameter, and initial leaf area for any of the treated plants prior to healing. This indicated that although wet mass and dry mass were only taken from a subset of plants in repetition 2 the same data in repetition would have identical results for wet mass and dry mass. The plants were very similar for each treatment within each repetition and across repetition as we saw no interactions between repetitions and treatments.

Final dry mass, Figure 4, was similar for all treatments with CWF 100 being significantly more than the B30:G2:R68 treatment. Mean plant compactness, Figure 5, showed that treatment CWF 100 was significantly more compact than B30:G2:R68 and B4:G5:R89 treatments. Figure 6, mean percent difference in leaf area showed no significant differences between treatments. Mean percent difference in height, Figure 7, showed that all but one treatment was statistically similar. The exception to this was the B4:G5:R89 treatment. All of the data between repetitions was combined due to the lack of statistical interaction among the two repetitions.

Conclusion

CWF 100 was significantly different from B30:G2:R68 for mean final dry mass resulting in a 21% increase in final dry mass. Light treatment had no effect on percent difference for leaf area (mm^2). Plant compactness (g/mm) calculation showed that treatments of CWF 100 significantly more compact, 22%, than 30B:G2:R68 and B4:G5:R89 treatments and similar to 50B:50R, 76B:24R, and CWF 75. B4:G5:R89 treatment had a 31% increase in percent difference in height and this was significantly different from all other treatments. Based on the results LED lighting can heal grafted tomatoes as well as Cool White Fluorescent lighting in a sole source lighting scenario.

Table 1.

Measurement	Units	Rep 1	Rep 2	Specifications
Avg. DLI	$mol \cdot m^{-2} \cdot d^{-1}$	8.4	22.2	One location at canopy height
Avg photoperiod	Hours	14.5	14.3	Average day length over 14 days; no supplemental lighting was used
Greenhouse air temp	°C Day	25.3±1.4	26.6±1.6	One location in the center of the greenhouse
	°C Night	25.5±1.4	27.5±1.8	
Greenhouse relative humidity	Day	85.1±7.5	82.8±5.0	Same location as greenhouse air temperature
	Night	81.8±8.5	80.0±6.3	
Greenhosue VPD	kPa Day	0.49±0.29	0.61±0.2	One location; calculated based on air temperature and humidity data
	kPa Night	0.61±0.32	0.75±0.28	
Atmospheric CO_2	$mmol \cdot mol^{-1}$	Ambient	Ambient	Natural ventilation
Nutrient solution pH		6.29±0.12	6.32±0.09	Data captured for each fertigation
Nutrient solution EC	$dS \cdot m^{-1}$	1.79±0.05	1.88±0.08	Data captured for each fertigation

Table 2.

	Temperature (°C)					
	30B:2G:68R	76B:24R	4B:5G:89R	50B:50R	CWF 75	CWF 100
Rep 1	27.93±0.45	27.6±0.50	27.92±0.34	27.76±0.37	27.4±0.39	28.01±0.28
Rep 2	27.81±0.33	28.01±0.33	27.84±0.29	27.94±0.22	27.8±0.29	27.84±0.29

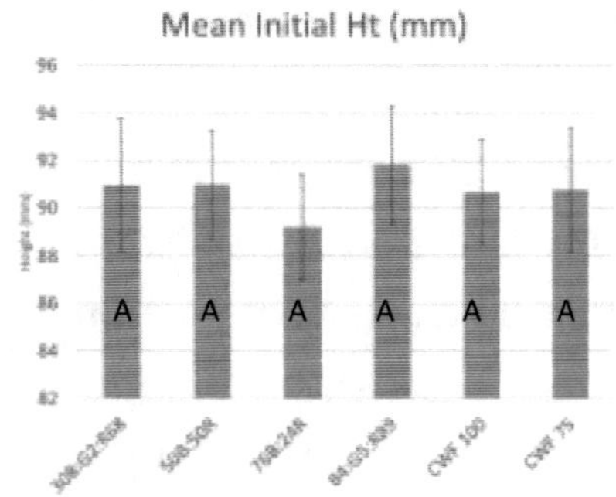

Figure 1.

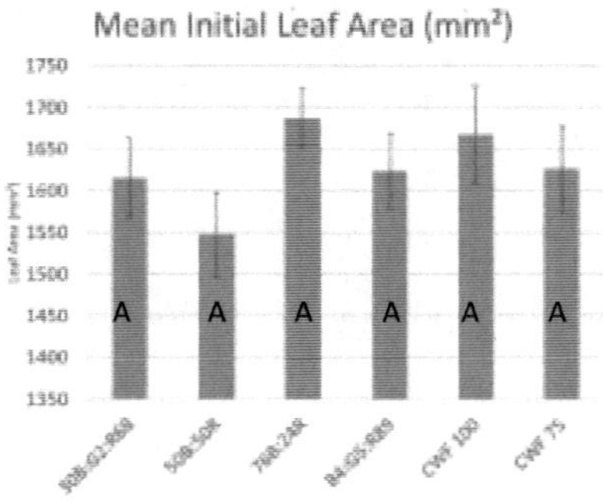

Figure 2.

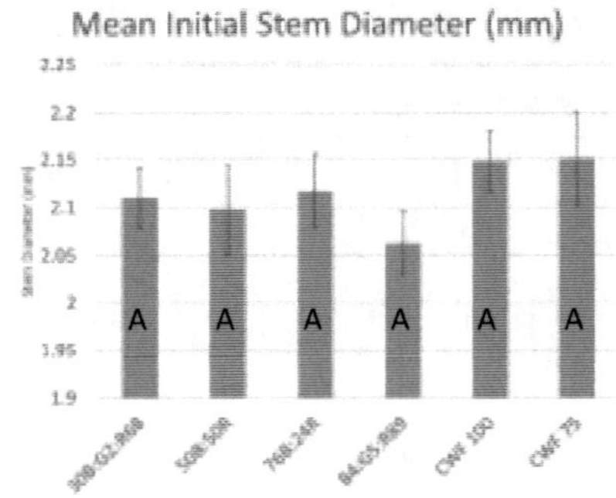

Figure 3.

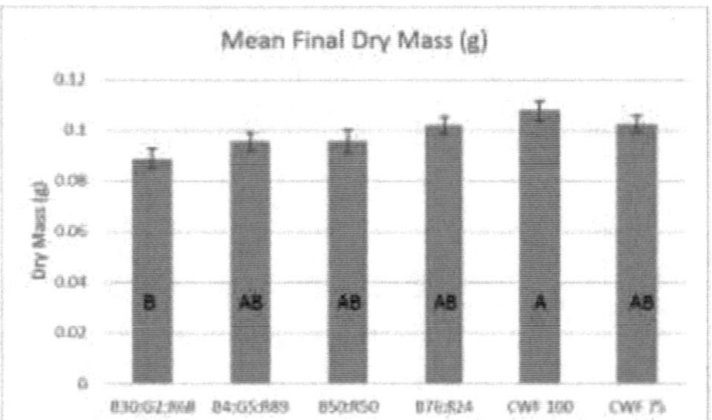

Figure 4.

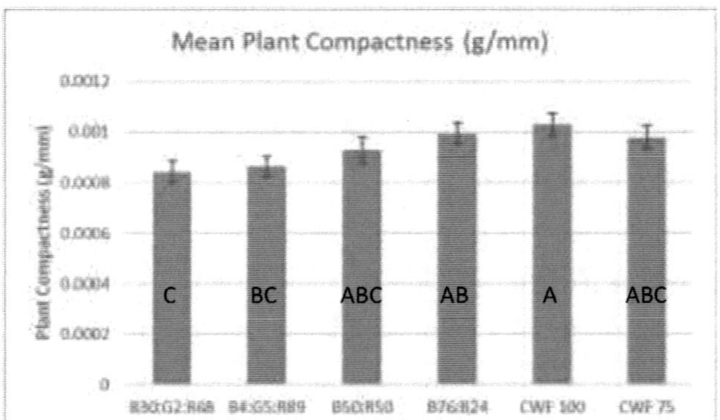

Figure 5.

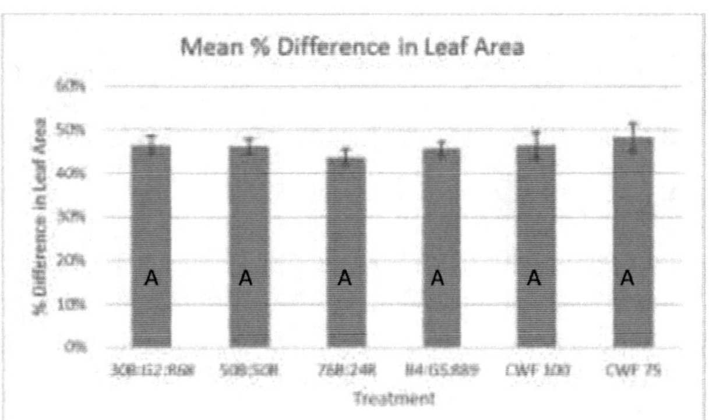

Figure 6.

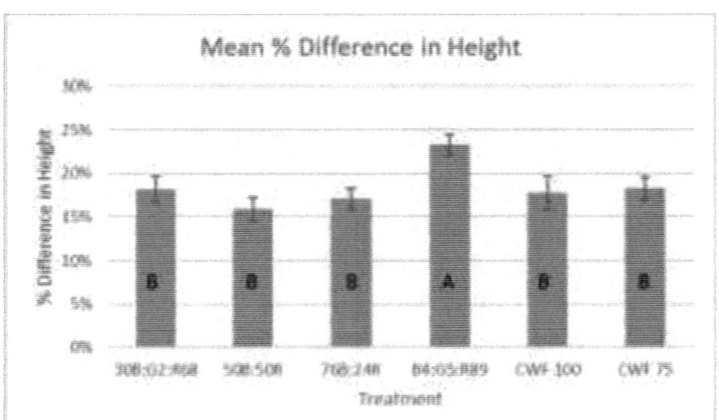

Figure 7.

PLANT GROWTH-PROMOTING BACTERIA: A SOLUTION FOR FOOD PRODUCTION IN THE WORLD

Bernard R. Glick
Department of Biology, University of Waterloo, Waterloo, Ontario, Canada N2L 3G1
E-mail: glick@uwaterloo.ca

In the future, in order to feed all of the world's people, it will be necessary to sustainably increase agricultural productivity. While several approaches to this problem will likely be needed, one way to do this is through the increased use of plant growth-promoting bacteria (PGPB). To help make this a reality, scientists need to develop a more profound understanding of the mechanisms employed by PGPB to facilitate plant growth. A key component of the efficacious functioning of PGPB to promote plant growth is the ability of the enzyme 1-aminocyclopropane-1-carboxylate (ACC) deaminase to lower plant ethylene levels. Moreover, the optimal functioning of PGPB includes the synergistic interaction between ACC deaminase and both plant and bacterial auxin, indole-3-acetic acid (IAA). PGPB that possess these traits not only directly promote plant growth, they also protect plants against various stresses including flooding, drought, salt, flower wilting, metals, organic contaminants, and both bacterial and fungal pathogens. While a considerable amount of both basic and applied work remains to be done before ACC deaminase-producing and IAA synthesizing PGPB become a mainstay of plant agriculture, the evidence indicates that with the expected future change from chemicals to soil bacteria, the world is on the verge of a major paradigm shift in plant agriculture.

Can we continue to feed all of the world's people in the future?

At present, there are ~7 billion people in the world, up from ~2.5 billion in 1950, and it is predicted to rise to somewhere around 9.5 billion by 2050. Thus, the increase in the world population in around the next 35 years will be approximately equal to the total number of people in the world in 1950. Moreover, as a result of both increasing population and industrialization, it is predicted that by 2050 worldwide agricultural productivity may actually decrease from its current level.

Some possible solutions

First, and possibly an impossible task, it is necessary to limit increases to the world's population. Second, it is imperative that we decrease food wastage. This includes better delivery of food from farms to cities and, in more affluent countries, decreasing portion sizes. Third, we need to optimize agricultural productivity on what we now consider marginal land. Fourth, we need to increase the use of transgenic plants that present us with the opportunity for increased crop yields and crops with higher nutritional levels. Fifth, we need to decrease the use of many agricultural chemicals, and instead begin commercializing on a large scale the use of plant growth-promoting bacteria and plant growth-promoting fungi.

Plant growth-promoting bacteria

The interaction between soil bacteria and plants may either be beneficial, harmful or neutral for the plant. Some plant growth-promoting bacteria (PGPB) began to develop a mutually beneficial relationship with flowering plants around 80 - 100 million years ago, so that the deliberate use of PGPB in agriculture mimics what has already been successful in nature. A large number of PGPB are commonly found in the plant rhizosphere, i.e. the area immediately around the roots, reflecting the fact that plant roots commonly exude a significant fraction of the carbon that is fixed by photosynthesis, and provide these carbon compounds to soil microbes that use them as a food source. In addition, some PGPB are endophytic and enter into a plant and colonize a portion of the plant's interior without harming the plant; and use essentially the same mechanisms as rhizospheric bacteria to promote plant growth and development.

Mechanisms used by PGPB

PGPB may impact plant growth and development either directly or indirectly. Direct promotion of plant growth occurs when a bacterium either facilitates the acquisition of essential nutrient resources from the environment or modulates of level of hormones within a plant. Nutrient acquisition includes the uptake of nitrogen, phosphorus and iron from the environment. PGPB may synthesize and provide to plants one or more of the phytohormones auxin, cytokinin and gibberellin. In addition, some PGPB can lower levels of the phytohormone ethylene by synthesizing an enzyme, 1-aminocyclopropane-1-carboxylate (ACC) deaminase, that cleaves ACC, the immediate precursor of ethylene in all higher plants. Indirect promotion of plant growth occurs when a PGPB (sometimes called a biocontrol PGPB) prevents, or decreases, the damage to that might otherwise ensue as a consequence of infection of the plant with any one of various phytopathogens. These phytopathogens typically include various soil fungi and bacteria.

Ethylene and ACC deaminase

The gaseous hormone ethylene is involved in seed germination, tissue differentiation, formation of root and shoot primordia, root branching and elongation, lateral bud development, flowering initiation, anthocyanin synthesis, flower opening and senescence, fruit ripening and degreening, production of volatile organic compounds responsible for aroma formation in fruits, storage product hydrolysis, leaf senescence, and leaf and fruit abscission. Ethylene synthesis is often a consequence of the response of plants to various biotic and abiotic stresses (this is sometimes called stress ethylene synthesis) including phytopathogens, high salt, flooding, drought, heavy metals, temperature extremes, organic contaminants, nematodes, Rhizobia nodule formation, and mycorrhizae-plant interaction. In some instances, ethylene is stimulatory while in other instances it is inhibitory of plant growth and development.

Stress ethylene has been reported to both alleviate and exacerbate some of the effects of various biotic and abiotic stresses (with the most studied stress being phytopathogen infection). A model that endeavors to explain the contradictory effects of stress ethylene on plants has been proposed. In this model, there is an initial small peak of ethylene that occurs close in time, usually within a few hours, to the onset of the stress. The small ethylene peak is followed (generally one to three days later) by a (second) much larger ethylene peak. The magnitude of the first peak is typically only a small fraction (1-5%) of the magnitude of the second peak and is thought to initiate a protective response by the plant, including turning on of the transcription of some defense-related genes. The first ethylene peak consumes the existing pool of ACC within plant tissues. The second ethylene peak is so large that processes such as senescence, chlorosis and abscission are initiated, the overall effect of which is inhibitory to plant growth and survival. Therefore, a large portion of the damage that occurs to a plant is due to autocatalytic ethylene synthesis and is not directly from the stress per se. Thus, exogenous ethylene can increase the severity of an abiotic or biotic stress, and ethylene inhibitors (chemical or biological) can significantly decrease the severity of such a stress.

The enzyme ACC deaminase which can cleave ACC and form ammonia and alpha-ketobutyrate, thereby lowering plant ethylene levels, is commonly found in a wide range of soil bacteria as well as in some soil fungi. It is a pyridoxal phosphate requiring enzyme with a monomeric size of 35-45 kDa and an active form that is either dimeric or trimeric. The cytoplasmically-localized enzyme binds ACC only relatively weakly, with a Km typically in the range of 1-15 mM. The enzyme is normally present at a low constitutive level in bacteria; however, the synthesis of additional enzyme may be induced by the presence of nM concentrations of the substrate ACC.

Model of ACC deaminase action

PGPB must first efficiently colonize either plant roots (either internally or externally) or seeds. These bacteria generally synthesize and secrete IAA, some of which is taken up by the plant, and together with the IAA synthesized by the plant, can either stimulate plant growth or induce the transcription of the gene encoding the plant enzyme ACC synthase, which is responsible for the synthesis of ACC. The IAA also loosens plant root cell walls thereby increasing plant root exudation. Along with other small molecules, some of the plant's ACC is exuded from germinating seeds or plant roots. The exuded ACC is taken up by the bacteria, and converted by ACC deaminase to ammonia and α-ketobutyrate. In this case, bacteria that contain the enzyme ACC deaminase effectively act as a sink for plant ACC. Consequently, the amount of ACC within plant tissues and hence the amount of ethylene produced by the plant is lowered and does not reach levels that are inhibitory for plant growth. The main short-term visible effect of seed inoculation with ACC deaminase-producing bacteria and also a simple biological assay for this enzyme is the enhancement of plant seedling root length. The observed effects are a direct consequence of the enzyme ACC deaminase and the short-term growth promotion effects persist throughout the life of the plant and are readily visible in soil.

PGPB vs. transgenic plants

The question arises whether PGPB or transgenic plants expressing the enzyme ACC deaminase are more effective at promoting plant growth and development in the presence of various stresses. To this end, transgenic tomato, canola and tobacco plants that expressed ACC deaminase were constructed. These plants included an exogenous ACC deaminase gene under the transcriptional control of either a plant promoter that was expressed in all plant tissues, a promoter that was activated following stress conditions, or a root-specific promoter. By far, in all three plants and with several different types of stresses, the root-specific promoter yielded plants that were the most resistant to the various stresses. In addition, in all instances, the PGPB-treated plants were inhibited to a lesser extent than were the transgenic plants. This latter result is attributed to the fact that PGPB do more for plants than merely lower ethylene, i.e. they provide the plant with other compounds to facilitate growth and overcome stress.

How do ACC deaminase-containing PGPB change plant gene expression?

Plant mRNAs synthesized in response to both wild-type (+ACC deaminase) and mutant (−ACC deaminase) PGPB were examined using microarray technology. While a large number of plant genes had altered expression levels as a consequence of adding either the wild-type or mutant PGPB strains, two striking responses were observed. First, plants that were treated with the mutant PGPB (but not with the wild-type) displayed a significant increase in the synthesis of mRNAs encoding stress response genes such as heat shock proteins and cytochrome P450s. This is consistent with the presence of the PGPB being a low level stress to the plant and the subsequent lowering of stress ethylene in the presence but not in the absence of ACC deaminase. Second, plants that were treated with the wild-type (but not with the mutant) displayed a significant increase in the synthesis of mRNAs encoding several auxin signal transduction factors (such as auxin transport and auxin response factors).

Revised model

The above mentioned data was incorporated into a revised model of PGPB functioning to better understand how these organisms promote plant growth (Fig. 1). Plant stress may increase the level of plant IAA production, the transcription of the ACC synthase gene (this enzyme converts S-adensoyl methionine, SAM, to ACC), and/or the transcription of the ACC oxidase gene (this enzyme converts ACC to ethylene). Ethylene induces the synthesis of a number of stress-inducible proteins. As it increases in the plant, ethylene also feedback inhibits auxin signal transduction (here the plant is trying to limit the synthesis of additional ethylene). In the presence of ACC deaminase, plant ACC is taken up by the bacterium and broken down thereby preventing a rise in plant ethylene levels with the result that i) plant stress proteins are not synthesized and ii) auxin signal transduction is not impeded. Thus, ACC deaminase, by lowering plant ethylene levels, facilitates the promotion of plant growth by increasing the flux of the phytohormone auxin (IAA).

Future prospects

As a consequence of recent developments in better understanding the fundamental mechanisms that are utilized by PGPB to promote plant growth, this technology is currently at a stage where it is amenable for implementation in agricultural, horticultural and environmental cleanup technologies in both the developed and the developing world.

References

Glick, B.R. 1995. The enhancement of plant growth by free-living bacteria. Can. J. Microbiol. 41: 109-117.

Glick, B.R., Penrose, D.M. and Li, J. 1998. A model for the lowering of plant ethylene concentrations by plant growth promoting bacteria. J. Theor. Biol. 190: 63-68.

Robison, M.M., Griffith, M., Pauls, K.P. and Glick, B.R. 2001. Dual role of ethylene in susceptibility of tomato to Verticillium wilt. J. Phytopathol. 149: 385-388.

Penrose, D.M. and Glick, B.R. 2003. Methods for isolating and characterizing ACC deaminase-containing plant growth-promoting rhizobacteria. Physiol. Plant 118: 10-15.

Glick, B.R. 2003. Phytoremediation: Synergistic use of plants and bacteria to clean up the environment. Biotechnol. Adv. 21: 383-393.

Glick, B.R. 2004. Bacterial ACC deaminase and the alleviation of plant stress. Adv. Appl. Microbiol., 56: 291-312.

Glick, B.R., Cheng, Z., Czarny, J. and Duan, J. 2007. Promotion of plant growth by ACC deaminase-containing soil bacteria. Eur. J. Plant Pathol. 119: 329-339.

Glick, B.R., Todorovic, B., Czarny, J., Cheng, Z., Duan, J. and McConkey, B. 2007. Promotion of plant growth by bacterial ACC deaminase. Crit. Rev. Plant Sci. 26: 227-242.

Glick, B.R. 2010. Using soil bacteria to facilitate phytoremediation. Biotechnol. Adv. 28: 367-374.

Glick, B.R. 2012. Plant growth-promoting bacteria: mechanisms and applications, Scientifica vol. 2012, Article ID 963401, 15 pages. doi:10.6064/2012/963401.

Duan, J., Jiang, W., Cheng, Z., Heikkila, J.J. and Glick, B.R. 2013. The complete genome sequence of the plant growth-promoting bacterium Pseudomonas putida UW4. PLoS ONE 8(3): e58640, 19 pages.

Reed, M.L.E. and Glick, B.R. 2013. Applications of plant growth-promoting bacteria for plant and soil systems. In: "Applications of Microbial Engineering". Gupta, V.K., Schmoll, M., Maki, M., Tuohy, M., and Mazutti, M.A. (eds.). pp. 181-229, Taylor and Francis, Enfield, Connecticut.

Glick, B.R. 2014. Bacteria with ACC deaminase can promote plant growth and help to feed the world. Microbiol. Res. 169:30-39.

Duca, D., Lorv, J., Patten, C.L., Rose, D. and Glick, B.R. 2014. Microbial indole-3-acetic acid and plant growth. Anton. Van Leeuwenhoek 106:85-125.

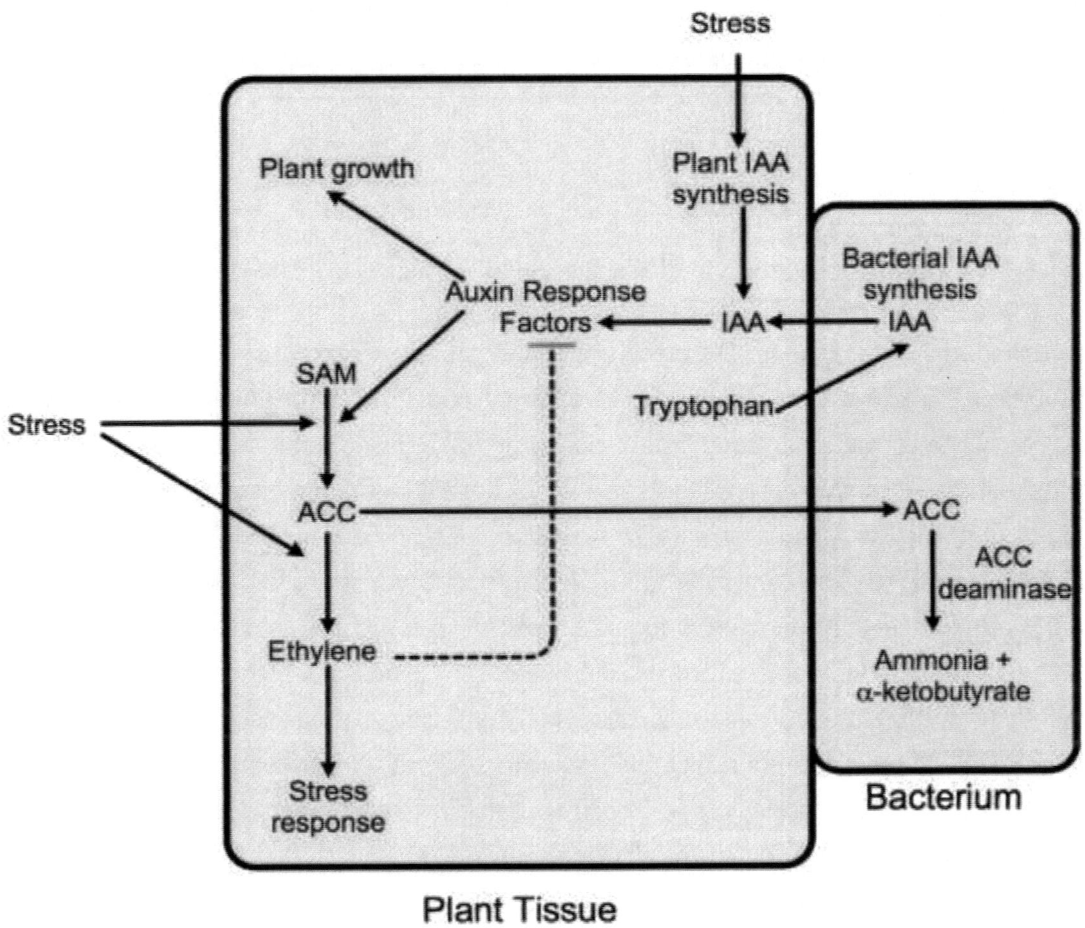

Fig. 1 How PGPB that produce both IAA and ACC deaminase promote plant growth.

IMPACT OF VARIOUS FORMULANTS, PLANT GROWTH REGULATORS, FUNGICIDES, AND COBALT ON ETHYLENE FORMATION IN SEEDLINGS OF DIFFERENT CROP PLANTS

Wilhelm Rademacher
BASF Global Research Crop Protection (retired), 67117 Limburgerhof, Germany

Solvents and surfactants, which are often used to formulate crop protection products, distinct plant growth regulators and fungicides, and different cobalt preparations were examined for their impact on ethylene formation in seedlings of barley, maize, oilseed rape, and soybean. The organic solvents tested [methanol, ethanol, acetone, dimethyl sulfoxide, Solvesso® 150 (an aromatic fluid)] turned out to be largely without effect. Both reduced [e.g., by Lutensol® AO 8 (an alcohol ethoxylate)] and increased [e.g. by Aerosol® OT-A (sodium dioctyl sulfosuccinate with 30% Solvesso® 150)] ethylene formation were found as a result of treatment with different surfactants. Aviglycine, salicylic acid, acetylsalicylate, prohexadione calcium, trinexapac ethyl, and pyraclostrobin inhibited ethylene formation in all four plant species. Activity varied among species with aviglycine generally being the most active compound. Cobalt treatments were most effective in soybean. Out of three tested cobalt preparations, Keylate® Cobalt turned out to be the most active inhibitor. - The obtained results indicate that (i) different plant species may react differently to different inhibitors of ethylene biosynthesis and that (ii) formulants may interfere both in a positive and in a negative sense with ethylene formation.

NATURAL AND TRINEXAPAC-ETHYL-INDUCED DROUGHT RESISTANCE IN WILLOW TREES (*SALIX ACMOPHYLLA*)

Joshua D. Klein[1], Maoz Aizikovitch[1], Yonit Raz Shalev[1] and Shlomo Cohen[2]
[1]Department of Natural Resources, ARO-Volcani Center, Rishon LeZion, Israel
[2]Institute for Agriculture according to the Torah, Yad Binyamin, Israel

The leaves of most foliage plants used as cut branches in the horticultural industry have thick cuticular and epicuticular layers, or low stomatal conductance rates, and thus can maintain turgor naturally for extended ornamental use. Willow trees usually require a steady supply of water, and their leaves wilt readily even under mild drought. Cut branches of willow that are used as part of Jewish ritual are sold in special plastic bags that do not always maintain foliage quality. Moddus (Syngenta brand of trinexapac-ethyl; TE) sprays of 0.5 or 1.0% (v/v) greatly reduced wilting of willow leaves in cut branches even three months after application. The relative water content of leaves on droughted branches was increased by as much as 100% as a result of TE treatment in comparison to untreated controls, while stomatal conductance was lowered by 50-70% and leaf and epicuticular wax thickness was increased by 30-50%. Two weeks after TE treatment, the leaves and bark in red-barked willows produced more salicylic acid, carotenoids, and anthocyanin, and had greater antioxidant activity, than corresponding tissues in green-barked trees of the same species. These metabolic effects were more pronounced in younger tissues, and were less evident three months after treatment. In addition to the horticultural benefit, the enhanced metabolites in TE-treated willows may provide a nutritional supplement for grazing goats and sheep.

PHOSPHORUS RESTRICTION AS AN ALTERNATIVE METHOD OF GROWTH CONTROL FOR PETUNIA

Josh B. Henry, Brian E. Whipker and Ingram McCall
Department of Horticultural Science, North Carolina State University, Raleigh, NC 27606 U.S.A.

Restricting the concentration of phosphorus (P) used in crop fertilization has a significant effect on plant growth. Low P concentrations are responsible for compact growth and shorter internodes in plants without the use of PGR's. Using lower P concentrations could potentially help growers to produce more compact plants while reducing fertilizer and chemical PGR expenditures. The objective of this study was to determine what P concentrations could be used to adequately control plant growth in two varieties of Petunia without the development of P deficiency symptoms. The variety 'Potunia Neon' represented a Petunia with a naturally compact growth habit, and 'Surprise Sky Blue' represented a more vigorous, trailing growth habit. Eight different P concentrations were used (0, 1.25, 2.5, 5, 10, 20, 40, and 80 mg·L^{-1} elemental P), and plants were grown using these concentrations for eight weeks. Tissue dry mass peaked at 10 mg·L^{-1} P while stem length peaked at 40 mg·L^{-1} P. Concentrations less than 5 mg·L^{-1} P developed P deficiency symptoms including delayed and limited flowering. Symptoms at low P concentrations were less apparent in the vigorous variety than in the compact variety, indicating a slightly higher P requirement for less vigorous varieties. The results indicate that P concentrations between 5 and 20 mg·L^{-1} can support healthy, compact growth for greenhouse grown petunias.

Introduction

For many years, it was believed that ammoniacal sources of nitrogen were the cause of plant stretch. It was not until Nelson et al. (2012) found that it is actually the concentration of phosphorus (P) that controls internode stretch, and the ultimate length of the stem. This knowledge may be used to aid growers in producing healthy and compact plants by restricting the amount of P that is applied to a crop.

To determine optimal concentrations of P for growth control in petunia (Petunia x hybrida), two cultivars were grown using a wide range of P concentrations. 'Surprise Sky Blue' was used to represent a vigorous growth habit, while 'Potunia Neon' was used to represent a compact growth habit.

Materials and Methods

Petunia cuttings were obtained from Dummen Orange, Inc. (Columbus, OH) on September 28, 2015 and rooted in 128 plug trays with cell dimensions of 2.7 x 2.7 x 3.8 cm. The substrate used was an 80:20 (v:v) mix of Canadian sphagnum peat moss (Conrad Fafard, Agawam, MA) and perlite (S & B Industrial Minerals North America, Inc, Morganton, NC), with added dolomitic lime at 251.3 g per cubic foot (Rockydale Agricultural, Roanoke, VA) and AquaGro 2000-G Wetting Agent at 17 g per cubic foot (Aquatrols, Cherry Hill, NJ). The plants were rooted under mist, and were given no fertilizer until treatments began. After the cuttings had rooted out, they were transplanted into 5-inch-diameter pots with dimensions of 12.7 x 9.7 x 9.0 cm. Potted petunias were grown in a glass greenhouse in Raleigh, NC.

The experiment was completely randomized with eight single plant replicates of each of the eight treatments. These eight treatments varied by P concentration, which was supplied in the form of potassium phosphate (KH_2PO_4). Nitrogen (N) and potassium (K) were held constant at 150 $mg{\cdot}L^{-1}$, with all other essential microelements remaining constant. Magnesium and sulfur varied slightly as six different fertilizer salts were used to maintain constant concentrations of N and K (Table 1). All of the nitrogen was supplied from nitrate-nitrogen (NO_3^-). Fertilizer solution was mixed in 100 L barrels, and was applied at each irrigation through 1.9 cm black irrigation tubing fitted with drip rings.

After eight weeks of growth, a destructive harvest was completed on both cultivars on December 17, 2015. During the harvest, plant diameters were recorded by measuring both the widest point and the plant turned at 90°, then averaging the two measurements. Substrate pH and electrical conductivity (EC) were obtained using the Pour-Thru method, and recorded using a Hanna HI 9813-6 portable meter. After these data were collected, dry mass was determined after tissue was allowed to dry for 72 hours at 70 °C.

Data for plant diameter, plant dry mass, substrate pH and substrate EC were analyzed using SAS (version 9.4; SAS Institute, Cary, NC). Data were subjected to PROC GLM and the means were separated by least significant differences at $P \leq 0.05$. PROC REG was then used to regress the data and determine the best fit linear or quadratic model for the different P concentrations.

Results

For 'Surprise Sky Blue', plants produced with 40 $mg{\cdot}L^{-1}$ P resulted in the largest dry mass and diameter. Plants produced with higher or lower concentrations of P were significantly smaller in both parameters, indicating peak growth at 40 $mg{\cdot}L^{-1}$ P. The 80 $mg{\cdot}L^{-1}$ P treatment was indifferent from concentrations of 5 – 20 $mg{\cdot}L^{-1}$ P in diameter, which all produced the second largest plants. Concentrations of 10 – 20 $mg{\cdot}L^{-1}$ P and 80 $mg{\cdot}L^{-1}$ P produced the second largest plants in terms of dry weight (Fig. 1). Substrate pH also followed an increasing trend up to a peak at 40 $mg{\cdot}L^{-1}$ P. The EC followed an inverse trend, due to the basic nature of the fertilizer. This trend was not different between the two cultivars. Plants that took up more salts from the substrate led to higher substrate pH.

Concentrations less than 5 mg·L^{-1} P resulted in plants exhibiting P deficiency symptoms including severe stunting, a dull grey coloration of the foliage and delayed flowering. Plants grown with no P did not flower at all. Additionally, plants produced with 0 and 1.25 mg·L^{-1} P displayed symptoms of lower leaf chlorosis and necrosis (Fig. 2). The second cultivar, 'Potunia Neon' exhibited similar patterns of growth with some differences. In this case, the 40 mg·L^{-1} P was not significantly different from 80 mg·L^{-1} P, with both concentrations producing the largest plants. Plants grown with P concentrations between 5 – 20 mg·L^{-1} had no significant differences, and were the next largest in both parameters. Substrate pH followed a similar increasing trend to 'Surprise Sky Blue', which peaked between 10 and 40 mg·L^{-1} P (Fig. 1). Concentrations less than 5 mg·L^{-1} P resulted in the same deficiency symptoms as 'Surprise Sky Blue' (Fig. 2).

Discussion

The overall size of the petunias grown in this study was very responsive to the concentration of P applied in the fertilizer solution. For both cultivars, concentrations of 40 mg·L^{-1} P resulted in the statistically largest plants, and the next largest plants by diameter were grown at concentrations between 5 – 20 mg·L^{-1} P. As there were no differences within this range of concentrations, one could use the lowest concentration of 5 mg·L^{-1} P to achieve the same effects.

Phosphorus concentrations less than 5 mg·L^{-1} resulted in deficiency symptoms that negatively affected the overall quality of the plant. This indicates that constant feed fertilization programs may go as low as 5 mg·L^{-1} P to control growth, and still produce an acceptable plant. A number of commercial fertilizers can be used by growers to supply this range of 5 – 20 mg·L^{-1} P. For instance, 13-2-13 Cal Mag mixed at 200 mg·L^{-1} N, 15-5-15 Cal Mag mixed at 100-200 mg·L^{-1} N, or 20-10-20 mixed at 100 mg·L^{-1} N could all be used to supply this amount of P to a crop.

We would like to express our gratitude to the Floriculture and Nursery Research Initiative for providing funds for this project.

Literature Cited

Nelson, P.V., S. CheonYoung, H. JinSheng, C.E. Niedziela J., and W.H. Swallow. 2012. Relative effects of fertilizer nitrogen form and phosphate level on control of bedding plant seedling growth. HortScience 47(2):249-253.

Table 1. Corresponding macronutrient concentrations for each of eight different phosphorus fertilizer treatments ranging from 0 to 80 mg·L^{-1}.

	Phosphorus Concentration (mg·L^{-1})							
	0.0	1.25	2.5	5.0	10.0	20.0	40.0	80.0
	Corresponding Macronutrient Concentrations (mg·L^{-1})							
Nitrogen	150.0	150.0	150.0	150.0	150.0	150.0	150.0	150.0
Potassium	150.0	150.0	150.4	150.0	150.0	150.0	150.0	150.0
Calcium	128.6	128.6	128.6	128.6	128.6	128.6	128.6	128.6
Magnesium	50.0	50.0	50.0	50.0	53.2	54.2	62.1	77.8
Sulfur	58.7	58.0	57.5	56.1	57.7	53.9	53.9	53.9

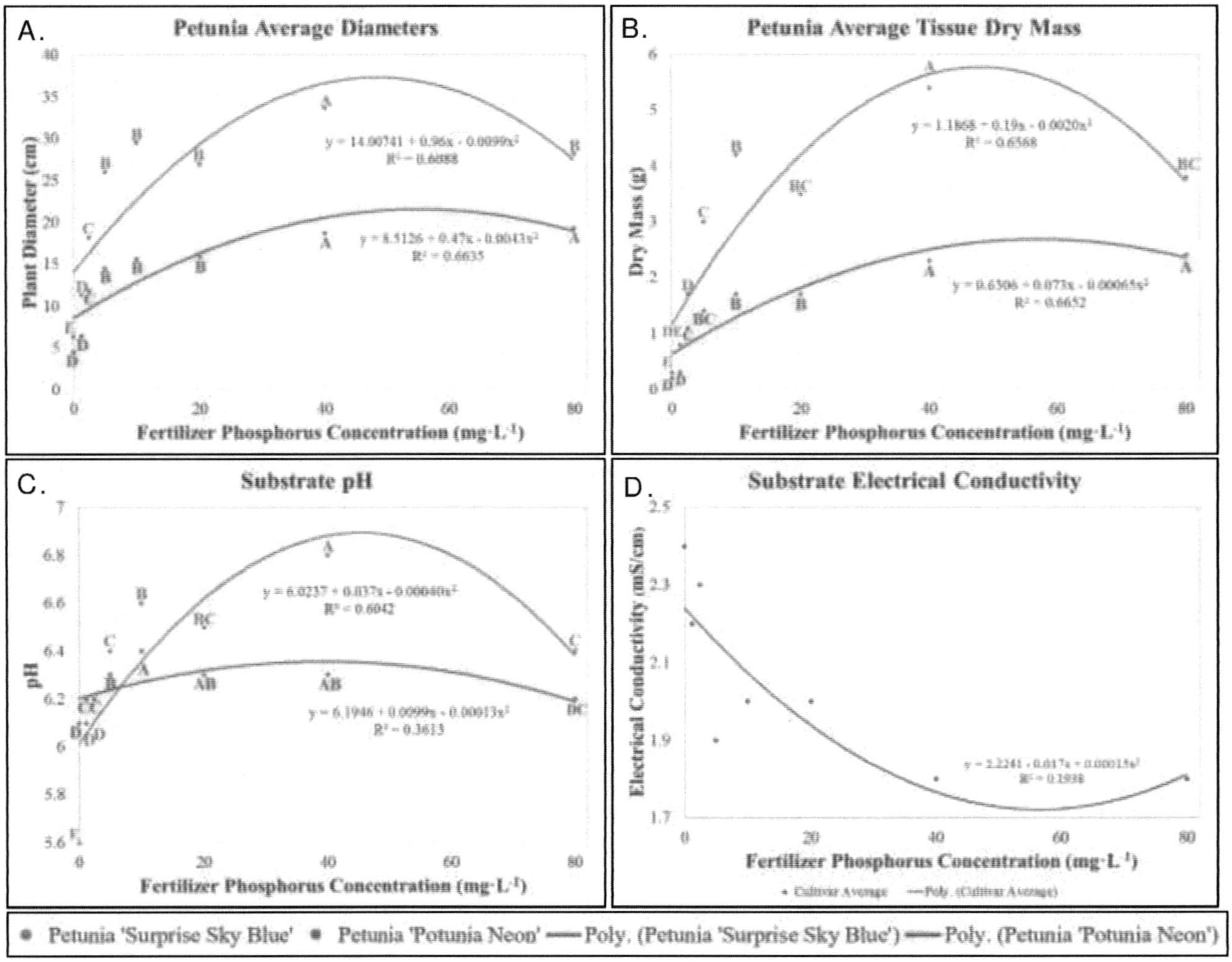

Figure 1. Growth of petunia 'Surprise Sky Blue' and 'Potunia Neon' at eight P concentrations: (A) average plant diameters, (B) average dry mass, and (C) average substrate pH for each cultivar separately. (D) Combined and averaged electrical conductivities for both cultivars.

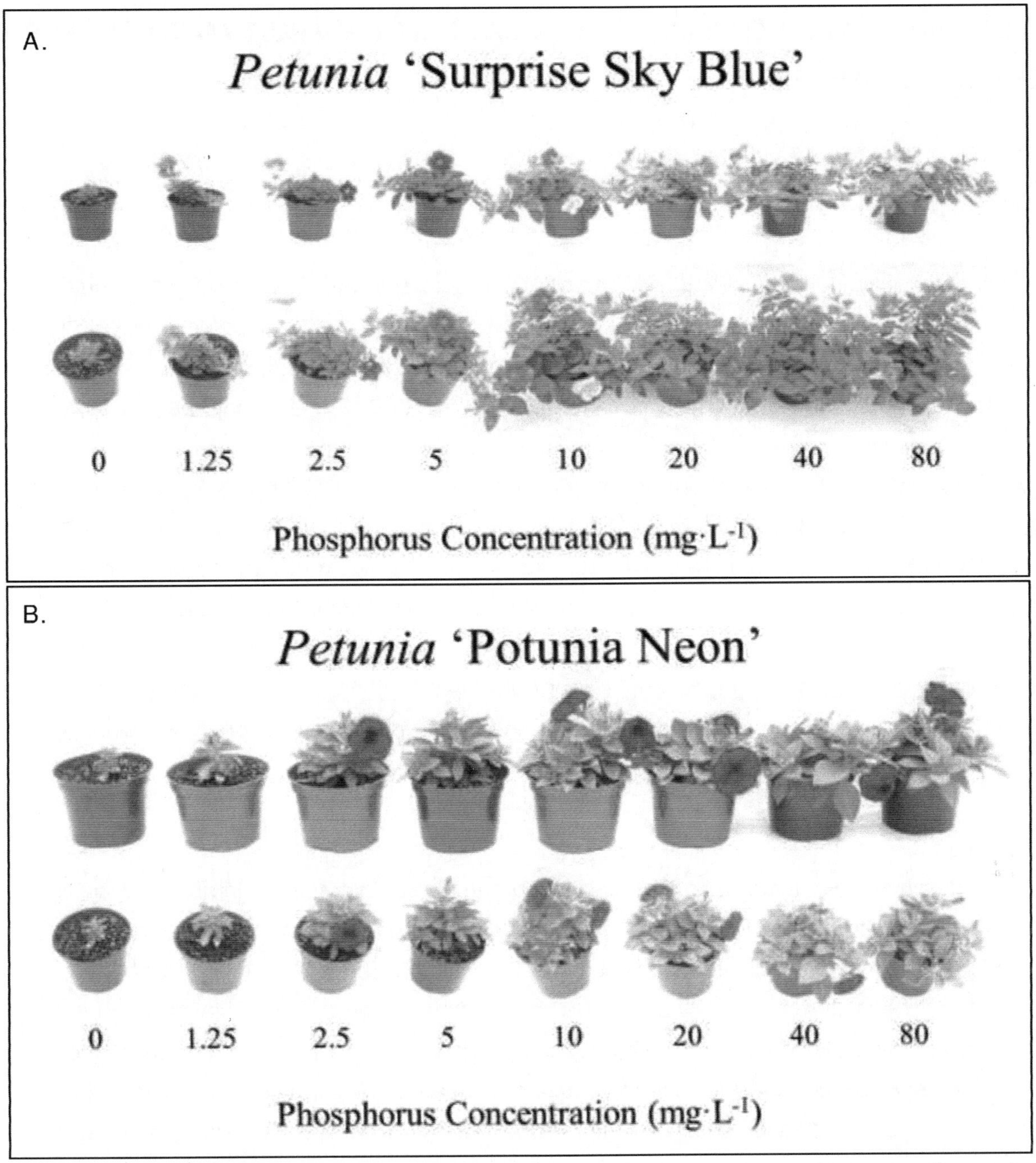

Figure 2. Petunia (A) 'Surprise Sky Blue' and (B) 'Potunia Neon' after eight weeks of growth using phosphorus fertilizer concentrations between 0 and 80 mg·L^{-1}.

AMINOETHOXYVINYLGLYCINE (AVG) PLUS GA3 DELAYED FRUIT MATURITY FOR A VERY LATE SEASON PEACH

Gregory L. Reighard and W. Greg Henderson
Clemson University, Clemson, SC 29634

Aminoethoxyvinylglycine (AVG) or Retain® in combination with ProGibb® was evaluated as a harvest aide to delay maturity, maintain or increase fruit firmness, and reduce pre-harvest drop on a drop prone, late season peach. Retain® (333g pouch) plus 591 ml of 4% ProGibb® in 378.5 l water with an organosilicone surfactant (0.05% v/v) was applied at 935 l/ha 14 days before harvest on 'September Sun' orchards in 2013 and 2104 in Ridge Spring and Monetta, South Carolina. Premature fruit drop with Retain® was marginally reduced in 2013 and 2014 in Ridge Spring, and in 2014 at Monetta. No differences in packout per tree in 2013 and 2014 at Ridge Spring were found with the treatment. No effect on marketable fruit size in the Titan packouts was noted for both years as 97% fruit exceeded 7 cm. Fruit from treated trees in the Monetta orchard had better overall size based on grower opinion, but fruit firmness was not increased in sampled fruit. Peak harvest maturity was delayed 2-3 days each year for both locations. A cost/benefit analysis of the Retain® + ProGibb® combination should be undertaken to determine its economic feasibility for closing gaps in the marketing season.

Introduction

Manipulating peach harvest maturity and increasing storage life can provide a significant advantage for marketing when gaps in supply occur due to weather events that result in a reduced crop for a specific time period. In addition, later season peach cultivars in the southeastern U.S. have a tendency to drop their fruit prematurely before harvest leading to a reduction in yields. High summer temperatures also can affect fruit firmness and negatively impact fruit quality throughout the distribution network before it reaches the consumer. Thus, a horticultural aid that can maintain fruit firmness, delay maturity and keep fruit from premature drop would prevent significant peach losses for late season cultivars.

Research on AVG (Retain®) with or without GA3 (ProGibb®) by Torrigiani et al. (2004), Cline (2006), and Chavez and Cook (2015) reported that AVG alone or in combination of these two products delayed fruit maturity and increased fruit firmness on mid to late season ripening peach cultivars and reduced preharvest drop in nectarines. However, the efficacy of this treatment on a very late season (September) peach cultivar that drops significant fruit numbers before harvest was unknown.

The objectives of this study were to maintain or increase fruit firmness and reduce pre-harvest drop on a drop prone, late season, peach cultivar and delay/extend the harvest season.

MATERIALS and METHODS

Two orchard locations of the cultivar September Sun, planted in 2009 (Titan Farms, 29 acres (11.6 ha), Ridge Spring, SC) and 2008 (Cotton Hope, 15 acres (6 ha), Monetta, SC), were set up as a CRB Design with 3 (Ridge Spring) or 2 (Monetta) replications for years 2013 and 2014.

The treatments were Retain® (aminoethoxyvinylglycine aka AVG) and ProGibb® (GA3) tank mixed and applied 14 days before harvest at 100 gals./acre (935 l/ha). Rates were one pouch of 333g (50g a.i.) of Retain® and 20 oz. (591 ml) of 4% ProGibb® per 100 gals. (378.5 l). An organosilicone surfactant was added at 0.05% (v/v). Treatments were applied early morning for a slower drying time.

Data collected included number of drops from 6 to 10 trees selected from each treatment in each rep counted before the first pick and 7 days later. Yield and packout (Titan Farms only) included total yields per tree, yields per date picked (maturity), fruit size, and % packout of number 1 fruit taken from packing line printout. Drop counts and fruit pressure/firmness of 20 randomly selected fruit measured with a penetrometer were recorded from the Monetta orchard.

Results

Premature fruit drop with Retain® was slightly reduced ($P < 0.10$) in 2013 in Ridge Spring and numerically in 2014 at both Ridge Spring and Monetta (data not shown). Retain® reduced drops by 28% in 2013 at Titan Farms and had a slight effect (<10%) on drops in 2014 at both sites (data not shown).

Yields per tree in 2013 and 2014 at Titan Farms for Retain® treated trees (71 and 76 kg, respectively) were similar (not significantly different) to the control trees (68 and 86 kg, respectively) (Tables 1, 2). Also, no significant differences in packout per tree in 2013 and 2014 at Ridge Spring were found with the treatment, and no effect on marketable fruit size in the Titan packouts was noted for both years as 97 % fruit exceeded 7 cm (data not shown). Fruit from treated trees in the Monetta orchard had better overall size based on grower opinion, but fruit firmness was not increased in the sampled fruit. Peak harvest maturity was delayed 2-3 days each year for both locations.

Conclusions

Retain® plus ProGibb® delayed peach maturity in a very late season peach cultivar in South Carolina. Reducing premature fruit drop was variable across years, but Retain® trees had less drop both years. A cost/benefit analysis of the Retain® + ProGibb® combination should be undertaken to determine its economic feasibility for reducing fruit drop and closing gaps and/or extending the marketing season.

Acknowledgements

Authors thank the staff at Titan Farms and Cotton Hope, and Valent BioSciences for their assistance and/or financial support for these trials.

Literature Cited

Chavez, D. and J. Cook. 2015. The effect of gibberellic acid (GA3) and aminoethoxyvinylglycine (AVG) on late-season peach [Prunus persica (L.) Batsch] varieties in Middle Georgia. HortSci. 50(9): S311-312.

Cline, J.A. 2006. Effect of aminoethoxyvinylglycine and surfactants on preharvest drop, maturity and fruit quality of two processing peach cultivars. HortScience 41:377-383.

Torrogiani, P., A.M. Bregili, V. Ziosi, S. Scaramagli, T. Ciriaci, A. Rasori, A. Biondi and G. Costa. 2004. Preharvest polyamine and aminoethoxyvinylglycine (AVG) applications modulate fruit ripening in 'Stark Red Gold' nectarines. Posthavest Biol. Technol. 33:293-308.

Table 1. Retain trial in 2013 with September Sun at Titan Farm, Ridge Spring, SC.

		Number of bins of packed fruit for each harvest date and percent of total in parenthesis.							Yield	Packout
Treatment	Rep	Aug. 26	Aug. 28	Aug. 30	Sept. 4	Sept. 10	Sept. 13	Total no.	(lbs/tree)	(%)
Control	1	3 (2.4)	16 (12.6)	29 (22.8)	49 (38.6)	25 (19.7)	5 (3.9)	127	164.2	73
Control	2	5 (4.0)	18 (14.5)	17 (13.7)	47 (37.9)	32 (25.5)	5 (4.0)	124	154.4	70
Control	3	3 (2.6)	24 (21.1)	19 (16.7)	42 (36.8)	22 (19.3)	4 (3.5)	114	138.8	87
Total no. (mean %)		11 (3.4)	58 (16.1)	65 (17.7)	138 (37.8)	79 (21.6)	14 (3.8)	365	152.5	76.7
Retain	1	5 (4.3)	22 (18.8)	21 (17.9)	28 (24.1)	36 (30.8)	5 (4.3)	117	142.4	69
Retain	2	2 (1.5)	17 (12.9)	22 (16.7)	39 (29.5)	44 (33.3)	8 (6.1)	132	173.3	74
Retain	3	3 (2.4)	20 (15.9)	26 (20.6)	45 (35.7)	25 (19.8)	7 (5.6)	126	165.3	72
Total no. (mean %)		10 (2.7)	59 (15.9)	69 (18.4)	112 (29.8)	105 (28.0)	20 (4.0)	375	160.3	71.7

Table 2. Retain trial in 2014 with September Sun at Titan Farm, Ridge Spring, SC.

		Number of bins of packed fruit for each harvest date and percent of total in parenthesis.							
Treatment	Rep	Sept. 2	Sept. 3	Sept. 4	Sept. 10	Sept. 16	Total no.	Yield (lbs/tree)	Packout (%)
Control	1	37.2 (34)	23.7 (22)	31 (28)	12.7 (12)	4.7 (4)	109	194	76
Control	2	57.4 (40)	30.4 (21)	38.4 (26)	14.2 (10)	4.6 (3)	145	199.7	78
Control	3	25.4 (33)	14.9 (19)	29.6 (38)	5.1 (6)	3 (4)	78	185.8	77
Total no. (mean %)		120 (36.4)	69 (20.9)	99 (30)	32 (9.7)	9.6 (2.9)	332	193.2	77
Retain	1	7.8 (11)	19.8 (29)	23.3 (34)	14.2 (21)	3.5 (5)	69	170.5	75
Retain	2	33.5 (29)	10.5 (9)	38 (33)	25.8 (22)	8 (7)	116	159.7	75
Retain	3	15.5 (20)	13.6 (17)	36.2 (46)	7.7 (10)	5 (6)	78	178.3	74
Total no. (mean %)		56.8 (21.6)	43.9 (16.7)	97.5 (37)	47.7 (18.2)	16.5 (6.3)	263	169.5	74.7

EFFECT OF STIMPLEX® ON YIELD AND GROWTH PERFORMANCE OF ORGANIC LEAFY GREENS IN TENNESSEE

Ravneet K. Sandhu and Dilip Nandwani
College of Agriculture, Human and Natural Sciences, Tennessee State University, Nashville, TN
Email: dnandwan@tnstate.edu

Stimplex® is the seaweed (Ascophyllum nodosum) extract used as a bio-stimulant known to have synergistic effect on the yield, growth and development of the plants. As in organic farming there are very limited resources of fertilization are available. Bio-stimulants provide nutrition and the alternates for the inorganic-synthetic fertilizers and beneficial as they are produced from algae. Organic farming is growing rapidly and becoming popular globally. Stimplex® is Organic Material Review Institute approved and safe to use in Organic farming. A study conducted at the certified organic field of Tennessee State University in Nashville. Two replicated research trials in spring 2015 and 2016 conducted. Six organic leafy green vegetables (lettuce, kale, collards, swiss-chard, mustard and amaranth) were treated with Stimplex® @2.5ml/gallon once a week for four weeks and yield performance recorded. Results from the study showed the significant difference ($p>0.05$) in the weight of the plants treated with Stimplex® and the untreated/control plants. In case of lettuce the head diameter measurement varied, though, no significant difference ($p<0.05$) in the values observed. Stimplex® treated plants were vigorous and robust than the control (untreated) plants. The per plant weight increase recorded in treated plants was, 33.2% in Mustard, 39.1% in Collards, 19.09% in Kale, 26.3% in Amaranth, 23.8% in Swiss-chard and 23.7% in Lettuce plants over control.

Introduction

Biostimulants are plant based extracts known to contain macro-elements, microelements, phyto-hormones, vitamins, sterols, betaines, polysaccharides etc., responsible for increase plant productivity. Alteration of phyto-hormones, improved photosynthetic efficiency, greater stomatal conductance, bio-synthetic regulation is known mechanisms that explain the Plant growth and development. Stimplex® is composed of seaweed extract (Ascophyllum nodosum) and is one of the most popular growth promoters in organic systems (Monica Popescu, 2013). Stimplex® is known to increase the yields of some vegetable crops and also enhance the firmness of fruits in organic systems (Fornes et. al, 1995 & 2002; Koo et. al, 1994; Noorie et. al, 2002). However, it is not clear if the Stimplex® would also affect the nutritional status in vegetable crops.

Due to the restriction on using synthetic fertilizers in organic systems, some organic bio-stimulators are applied as plant growth promoters. Studies showed that the use of bio-stimulants in small amounts have been effective on some metabolic activities, improved plant development and advancement by increasing the rate of photosynthesis, endogenous hormones, ionic uptake, synthesis of protein and in addition, with moderately higher capacity for intensifying micronutrients accessible in the soil (Snedecor et. al, 1980). Plant bio-stimulant refers to a material which contains substance(s) and/or microorganisms whose capacity, when applied to plants or the rhizosphere, is to fortify common methodologies giving advantage to supplement uptake, supplement proficiency, resilience to abiotic push, and/or crop quality, autonomously of its supplement content (Xu et. al., 2015). The worldwide business sector for bio-stimulants is anticipated to increment 12 % every year and reach over $2,200 million by 2018 (Pamela et. al., 2014).

Biostimulants have no direct activity against pests, and hence do not fall inside the administrative structure of pesticides (European Biostimulants Industry, 2012). Stimplex® is a bio stimulant derived from seaweed extracts. Stimplex® is used as fertigation through drip irrigation. It is applied after two weeks of emergence and thereafter at one-week intervals until all heads blossom (Alessandra et. al., 2014).

There are many studies that investigate the nutrition of organic vegetables compared to conventionally grown vegetables. However, there are not many studies or research on the nutritional determination of the bio stimulant treated organic vegetables. Biostimulants are relatively new products for use in organic crop cultivation as a growth promoter, but its effects on mineral content or nutrition of the plants is still unknown requiring further research.

Materials and Methods

Materials: Stimplex® (Acadian Sea-plants Limited, Canada), seeds (certified organic) of Lettuce, Kale, Mustard, Amaranth, Collards and Swiss-chard.

Methodology: This study was conducted in spring of 2015 and 2016 growing season at the certified organic field of Tennessee State University campus farm. Selected leafy greens viz. Lettuce, Kale, Swiss chard, Amaranths, Collards and Mustard greens were evaluated for growth and yield performance in an organic production system when treated with Stimplex®. The production practices during whole experiments in field were followed as per standards of National Organic Program (NOP). Stimplex® and seeds used for the evaluation are commercially available from seed companies (Johnny Seeds, Sow Seeds, Territorial Seed Company).

Certified organic or untreated seeds were planted in spring 2015 and repeated in 2016 growing seasons. Seeds were direct seeded into seedling trays or jiffy cups filled with an organic vermi-compost. Seedlings were watered as required and fertilized with a 120 mg/L nitrogen (N) solution (mighty grow- living fertilizer) and kept in the protected conditions (greenhouse). Vermi-compost used for seedling production was OMRI (Organic Materials Review Institute) listed. Thinning was done as seedlings grew. Seedlings were then transplanted in the certified organic field plots as they attain suitable height and vigor. The experimental design was complete randomized block (CRBD) with three replications. There were two rows of 6 plants in each row per plot for each crop, total of 216 plants in the field at the spacing of 1' and 1.5' between the plants and row to row, respectively. One row of plants was kept control and the other row of plants was treated with Stimplex® @2.5 ml/gallon. First application of Stimplex® was done after 10 days of transplanting. Total 5 applications were applied at weekly interval. Soil moisture data was collected by Soil Moisture Meter to determine the irrigation frequency. Drip irrigation system was used during the whole field trials. Before each trial, soil samples were analyzed for macronutrients. Land preparation and tillage was accomplished with a rotary digging machine and secondary tillage was done with an offset disc.

No other synthetic fungicides or bactericides were applied to the trials. Weeds were managed by hand cultivation or mechanical tools throughout the trial. Crop plants were evaluated during the trial for overall appearance and robustness. At maturity, six of the plants of each row in each plot were harvested. A second harvest was conducted on those crops wherever common production practices required multiple harvests (Coolong et al., 2013). In case of lettuce, each head was weighed for fresh weight and head diameter recorded. For other crops the leaves were harvested and per plant weight data was recorded.

Data from all seasons and years was analyzed using analysis of variance (ANOVA) (PROC GLM), and mean separation using LSD test ($P < 0.05$) with SAS (version 9.1). ANOVA included treatment and year and their interactions as sources of variance (Coolong et. al., 2013).

Results and Discussions

All the crops were harvested at respective maturity levels. The total yield in lbs./acre and per plant weight in g is represented in Table 1. All the treated plants in all crops showed increase in yield. However, this increase was not significant in case of some vegetables. Mustard plants treated with Stimplex® had more total yield (33003.78 lbs./a) and average yield per plant (515.42 g). However, this difference in yield of treated and control plants was not significant ($p > 0.05$). Leaf lamina of treated plants was also higher than the control plants. Lettuce plants treated with Stimplex® performed better and gave significantly ($p < 0.05$) higher yields than the control plants. Total yield per acre of treated and control plants were 21808.64 lbs. and 17616.24 lbs., respectively. There was significant difference in the head diameter values of treated and control lettuce plants.

Stimplex® treated Kale, Collard and Amaranth plants also had the higher yields and leaf lamina. However, this difference in the values of yield was not significant. Kale leaf lamina values were significantly ($p<0.05$) different from each other. Swiss-chard plants yield more when treated with stimplex® and the increase is significantly different. All the treated plants were more robust and vigorous as compared to control plants. The control plants were attacked by aphids and beetles; those eat the leaves of plants. Similar study done on cucumber plants. This study reported Stimplex® application leads to significant decrease in disease incidence of disease causing pathogens (Jayaraman et al., 2010)

The leaves of collard plants treated with stimplex® were more smooth, leathery and shiny than the control plants. Mustard control plants mature earlier viz. showed flowering earlier than the control plants. Leaves of Swiss-chard, Collards and Mustard plants treated with stimplex® were more dark in color than the control plants. Similar study on bio-stimulants reported that these products significantly increases the overall yield per tree, yield per hectare and fruit size of different cultivars of apples compared to the untreated or control trees (Dominguez et al., 2015). Study reported in Romania also concluded that Biostimulants are significant to increase the plant production (Monica Popescu, 2013).

Conclusion

All the treated plants of six crops showed the higher average per plant (g) and average total yield per acre (lbs.). However, some of crops these values are not significant. Moreover, the Stimplex® had very synergistic effect on the growth of the plants. Stimplex® can be a better alternative to the synthetic chemical fertilizers or organic fertilizers in organic agriculture. However, more studies on different dose delivery of Stimplex® needed.

Acknowledgement

Authors would like to thank Dr. Jeffrey P Norrie of Acadian Sea-plants Ltd. for providing stimplex. Senior author (RS) is grateful to PGRSA conference organizers for travel funding. Our sincere appreciation to Organic research program team in field trials. Funding for the research received from Evans-Allen grant (#222305) to DN.

References

Alessandra, T. A., Marcucci, M., Renzaglia, Elvira, Rahmann, G. and U. Aksoy. (2014). Filtrate seaweed extract as bio-stimulant in nursery organic horticulture. Organic eprints. Thuenen Report 20: 697-700.

Coolong, T., Derek, M. L., John, C. S., Brent, R. and M. A. Williams. (2013). Organic Leafy Greens Cultivar Trials in Kentucky: Identifying Superior Cultivars for Small-scale Organic Farmers. HortTechnology 23: 241-246.

European Biostimulants Industry Council (2012a) EBIC and biostimulants in brief. http://www.biostimulants.eu/.

Fornes, F., M. Sánchez-Perales, and J.L. Guardiola. 1995. Effect of a seaweed extract on citrus fruit maturation. Acta Hort. 379:75–82.

Fornes, F., M. Sánchez-Perales, and J.L. Guardiola. 2002. Effect of a seaweed extract on the productivity of 'de Nules' Clementine mandarin and Navelina orange. Botanica Marina 45:486–489.

G.W. Snedecor, G. W. Cochran. (1980) Statistical Methods. 7th Ed. The Iowa State Univ. press Iowa, Ames, USA.

Jayaraj Jayaraman, Jeff Norrie, Zamir K. Punja. (2010). Commercial extract from the brown seaweed Ascophyllum nodosum reduces fungal diseases in greenhouse cucumber. Journal of Applied Phycology, 2010, Page 1.

Koo, R.C.J. and S. Mayo. 1994. Effects of seaweed sprays on citrus fruit production. Proc. Fla. State Hort. Soc. 107:82–85.

Leonel I. Dominguez and Terence L. Robinson. (2015). Strategies to Improve Early Growth and Yield of Tall Spindle Apple Plantings. New York fruit quarterly. Volume 23. Number 2.

Monica Popescu. (2013). Agricultural Uses of Seaweeds Extracts. Current trends in Natural Sciences. Volume 2, Issue 4.

Norrie, J., T. Branson, and P.E. Keathley. 2002. Marine plant extracts impact on grape yield and quality. Acta-Horticulturae. 594:315–319.

Pamela Calvo, Louise Nelson, Joseph W. Kloepper. (2014). Agricultural uses of plant Biostimulants. Plant and Soil, Page1.

Xu, C. and I. L. Daniel (2015) Effects of A. nodosum seaweed extracts on spinach growth, physiology and nutrition value under drought stress. Scientia Horticulturae 183: 39–47.

Table 1. Average total and per plants yield of crops for spring 2015 and 2016.

Crops	Total yield per acre (lbs.)		Average per plant wt. (g)	
	Treated	Untreated	Treated	Untreated
Mustard	33003.78 [a]	24759.50 [a]	515.42	386.67
Lettuce	21808.64 [b]	17616.24 [c]	340.58	353.78
Kale	23338.32 [d]	17356.55 [d]	364.47	271.06
Collard	17819.02 [e]	12808.42 [e]	278.28	200.03
Swiss chard	16010.08 [f]	12927.59 [g]	250.03	201.89
Amaranths	8262.06 [h]	6540.28 [h]	135.39	102.14

Table 2. Average leaf lamina of plants grown in year 2015 and 2016

	Average leaf lamina (cm)	
Crop	Treated	Untreated
Mustard	27.3a	22.4b
Kale	14.5c	11.9d
Collard	13.0e	11.4e
Swiss Chard	13.2f	11.5f

INFLUENCE OF FOLIAR SPRAYING WITH MICROALGAL AND KELP EXTRACTS ON THE MINERAL NUTRITION AND YIELD OF MAIZE HYBRIDS KN 307 AND KN 442. OUTDOOR STUDY

Konstantin Chakalov[1], Valentin Savov[2], Gergana Angelova[1], and Todorka Popova[1]
[1] Balkan Plant Sciences ltd, 25A Tsar Osvoboditel blvd., Sofia, Sofia 1164, Bulgaria
[2] Sofia University "St.Kliment Ohridski", 8 "Dragan Tzankov" Blvd., 1164 Sofia, Bulgaria
E-mail: konstantinchakalov@yahoo.com

In our previous year's study we discussed the role of microalgal extracts as root accelerators. The phytohormonal composition of the extracts showed that they might be able to be competitors of some kelp products (Chakalov, K. et al. 2015). That and some other preliminary tests provoked the idea to test microalgal extracts as bioregulators in maize production.

The aim of this study is to determine the possibility of using locally produced algae as an alternative of kelps in the preparation of native PGRs for maize treatment.

Material and Methods

The impact of kelp extract Ascophyllum nodosum, microalgal extracts - Chlorella and Spirulina, and microalgal+ Moringa oleifera leaves extracts on the mineral nutrition and the yield of two maize hybrids was tested. The field experiment was conducted at the Maize Institute in Knezha with maize hybrids Kn 307 and Kn 442 under conditions without irrigation, on a typical slightly leached soil (Haplic Chernozems) and background of nutrition $N_{80}P_{100}$, in 3 replications and on a 10 m^2 plot. The tested preparations were obtained by a double enzyme extraction. The plants were treated in the 10th leaf phase with a dose of 500 ml/ha 0,4% solution from each preparation. Ear leaves for leaf diagnostics (mineral nutrition and pigments content) were gathered during the silking phase.

Results

Kelp extract treatment increased the N, P, and K content of the indicatory leaf in the flowering phase by 14.47, 10.78, and 11.87% for Kn 307, and by 8.09, 11.33, and 3.78 % for the more late-maturing Kn 442. Microalgal extracts had more moderate impact on the assimilation of those elements in the case of Kn 307, although the treatment with Spirulina extract increased the content of N, P and K in the indicatory leaf of Kn 442 respectively by 14.17, 16.75, and 16.82% (fig.3 and 4). The extracts influenced positively the chlorophyll a and b content of the tested hybrids. In the case of the early-maturing hybrid the maximum yield increase (9.2%) was achieved by treatment with Spirulina extract, and the content of proteins and lipids in the grain was increased by 14.05 and 16.19% respectively (fig. 1 and 2). The yield results for hybrid Kn 442 demonstrate that treatment with kelp and microalgal extracts doesn't result in statistically proven differences with the control, but the application of Ascophyllum nodosum extract or a mixture of Chlorella and Spirulina leads to an increase in fat content of 12.22 and 13.03% respectively (fig. 5 and 6). This field experiment shows that microalgal extracts are competitive to kelp extracts in the growing of early and medium-early maize hybrids.

Conclusion

Microalgal extracts are an acceptable alternative for including in the composition of PGRs for maize treatment instead of kelp extracts.

References

Chakalov Konstantin, Valentin Savov, Gergana Angelova, and Todorka Popova (2015) study of the effect of natural root accelerators on plant rooting. Annual meeting of PGRSA 2015

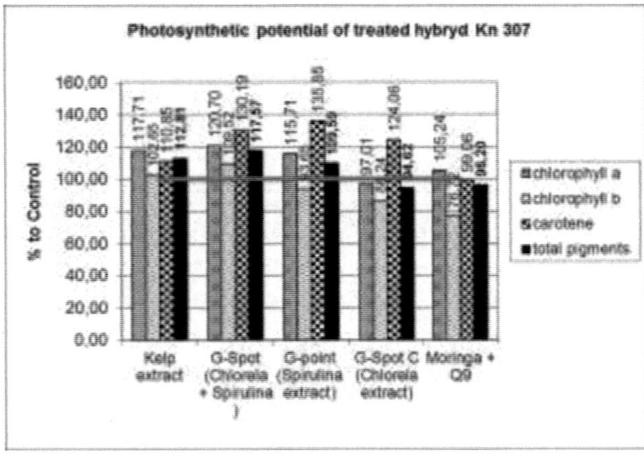

Figure 1.

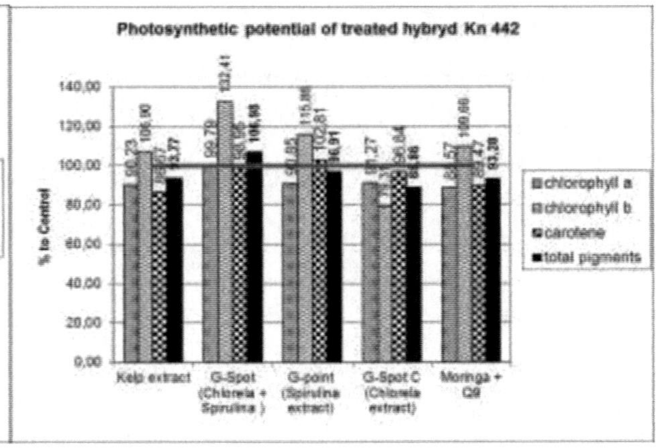

Figure 2.

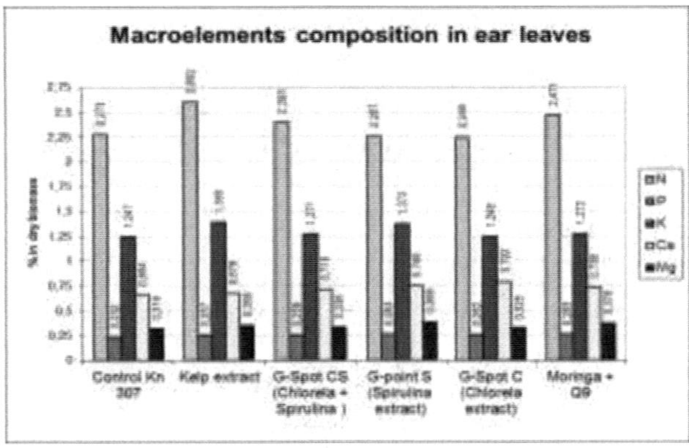

Figure 3.

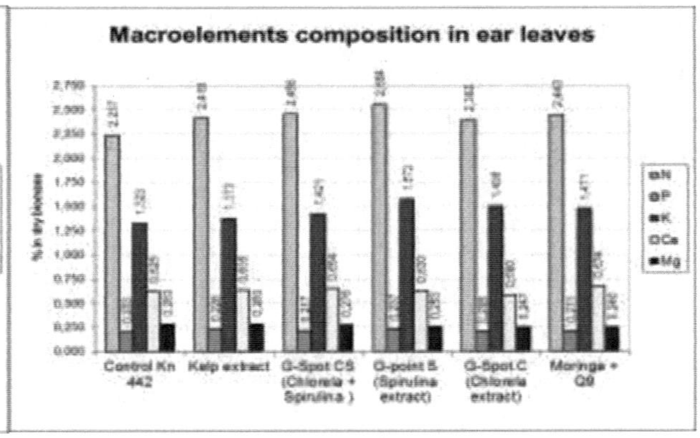

Figure 4.

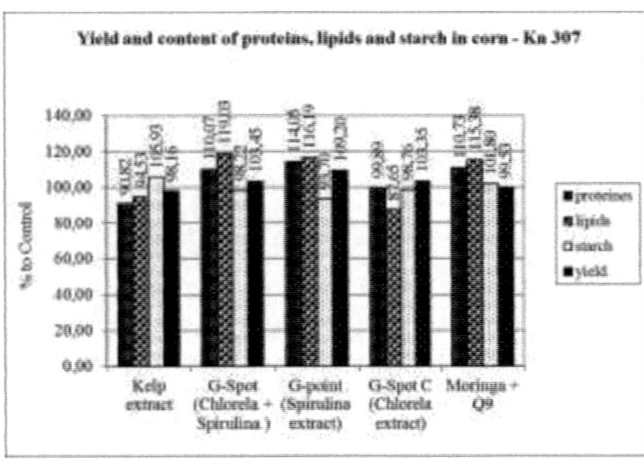

Figure 5.

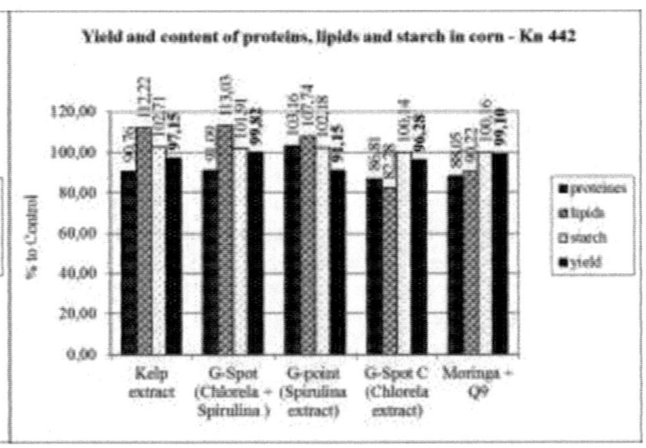

Figure 6.

GENE EXPRESSION DIFFERENCES IN HYDROGEN CYANAMIDE-TREATED PEACH TREES

Shweta Chhajed[1], T. Vashisth[1], T. Colquhoun[2], J. Olmstead[3] and M. Olmstead[3]
[1]Horticultural Sciences Department, University of Florida/Institute of Food and Agricultural Research, Citrus Research and Education Center, Lake Alfred, FL;
[2]Environmental Horticultural Department, University of Florida/Institute of Food and Agricultural Research, Gainesville, FL;
[3]Horticultural Sciences Department, University of Florida/Institute of Food and Agricultural Research, Gainesville, FL

Peaches (Prunus persica L.) are of considerable interest as an alternative crop for growers in Florida who are looking to diversify from citrus. The temperate origin of peach present significant production challenges for subtropical areas. Most significantly, inadequate chill accumulation is a common occurrence in production areas seeking to take advantage of high value, early-market windows. Low-chill peach varieties grown in Florida, USA that do not receive adequate chill exhibit extended bloom periods (>2 weeks in length). Accumulating inadequate chill units for uniform bloom is a major challenge, as it can result in high labor costs for multiple thinning and harvesting passes in the orchard. Most of Florida's chill hours are accumulated in December and January, but in recent years accumulated chilling units continue to decline. Hydrogen cyanamide (HCN) is widely used around the globe where mild winters prevail and where there are insufficient accumulated chill units to coordinate budbreak, induce uniform leaf emergence, and compress the bloom period. However, many growers have induced significant crop injury because the timing for different fruit crops is not well defined. Additionally, the molecular mechanism of how HCN works is still poorly understood and therefore it is challenging to develop strategies for proper and effective use HCN on fruit crops in absence of chill hour accumulation and erratic weather. In order to understand the transcriptional pathways associated with HCN-induced dormancy release in peaches we analyzed the vegetative and reproductive buds of peach. The objective of this study was to examine how HCN affects gene expression of key genes associated with endodormancy and budbreak in peach floral and vegetative buds grown in subtropical climates. On December, 19 2016, when pollen grains were translucent, low chill cultivar 'TropicBeauty' trees were sprayed with 1% HCN (BudPro) and control trees were not sprayed. Vegetative and floral buds were collected from 4 branches per tree for 4 biological replicates before application and then 1, 3 and 7 days after application. RNA isolation was performed to carry out cDNA synthesis for real time PCR (RT-PCR). Three genes involving dormancy release [DAM4 (dormancy-associated MADS-box transcription factor), DAM5, and DAM6] and one gene involved in flower bud initiation (AP1) were targeted for gene expression studies. HCN treated plants had both early and uniform bloom, vegetative bud break was visible approximately 30 days after HCN application. Whereas untreated control trees displayed a prolonged bloom and uneven leaf emergence suggesting HCN application helped in breaking the dormancy and synchronizing the bud break. Our gene expression results show that DAM 4, DAM5, and DAM6 were down-regulated after 7 days of HCN application. Previous literature suggests that DAM genes are related to dormancy in perennial trees. DAM genes are known to up-regulate during endodormancy and down-regulate with release of endodormancy. Altogether our field and gene expression data shows that HCN application can induce early and uniform bud break potentially by substituting for chill hours required for release of endodormancy and the transcriptional pathway is similar to innate endodormancy release in peach. To have a better understanding of transcriptional pathway associated with HCN-induced release of endodormancy we plan to look at genes involved in the vernalization pathway and investigate if they follow a similar pattern as that of peach trees that receive the natural chilling hours. Therefore, we will be investigating MADS-box gene, FLOWERING LOCUS C (FLC), which is a known key flowering repressor regulated by vernalization. FLC represses flowering in a protein complex with another MADS-box protein, SHORT VEGETATIVE PHASE (SVP), by binding to the regulatory sequences of flowering time genes SUPPRESSOR OF OVEREXPRESSION OF CONSTANS 1 (SOC1). This SOC1 gene and AP1 gene together activates LFY (Leafy) which turns on the expression of all the floral homeotic genes. Additional experiments with HCN application will be continued with these genes to validate our hypothesis.

CONTROLLING HEIGHT IN ECHINACEA HYBRIDS

Mara C. Grossman, Margaret Aiken, Holly Scoggins and Joyce Latimer
Department of Horticulture, Virginia Tech, Blacksburg, Virginia 24061

Hybrid cultivars of Echinacea purpurea (L.) Moench, eastern purple coneflower, vary significantly in their growth habits and production requirements from the species. Some Echinacea cultivars can grow to be very tall and leggy. Our objective was to examine the effects of plant growth regulators (PGRs) on height in two different cultivars of hybrid Echinacea. We wanted to determine if different concentrations and application methods were effective in controlling plant and flower height. Liners of Echinacea 'Marmalade' and 'Harvest Moon' were received in March 2015 and transplanted into trade gallon pots filled with peat/pine bark media (Fafard 3B, SunGro Horticulture, Vancouver, CA). Plants were kept under short days for six weeks. After short days ended, the following PGRs were applied as foliar sprays: Daminozide (Dazide, Fine Americas, Inc., Walnut Creek, CA) 5000 ppm (2 applications); Uniconazole (Concise, Fine Americas, Inc., Walnut Creek, CA) 30 ppm (1 application); Uniconazole (Concise) 15 ppm (2 applications); Paclobutrazol (Piccolo 10 XC, Fine Americas, Inc., Walnut Creek, CA) 80 ppm (1 application); Paclobutrazol (Piccolo 10 XC) 40 ppm (2 applications); or, as a single drench application (10 fl. oz. per pot): Uniconazole (Concise) 2 ppm or Paclobutrazol (Piccolo 10 XC) 4 ppm. All PGR treatments were compared to an untreated control. The second application was applied two weeks after the first. Plants were grown until they reached marketable size (4 weeks after initial treatment for 'Marmalade' and 5 weeks after initial treatment for 'Harvest Moon'). At this time plant height, flower stalk height, width, days to flower, number of flowers, market rating, and shoot dry weight were measured.

The effectiveness of height control PGRs varied by cultivar. 'Marmalade,' which has very vigorous growth, only had reduced height with daminozide; plants treated with daminozide were 25% shorter than untreated controls. However, daminozide had the unwanted effect of reducing the number of branches and flowers in 'Marmalade,' 6.2 branches compared to 9.8 in controls, and 3 flowers compared to 10 in controls. All PGR treatments reduced shoot weight in 'Marmalade,' with the greatest reduction seen in plants treated with daminozide (17.7 grams compared to controls 44.0 grams).With this cultivar, possibly earlier application of height control PGRs would be more effective; more work will need to be done to determine optimal timing. 'Harvest Moon,' which has a more compact growth habit compared to 'Marmalade,' had decreased height in response to all of the height control PGRs applied, with heights in treated plants reduced from 25-41% compared to controls. The best results in 'Harvest Moon' were seen in plants treated with uniconazole 30 ppm spray and paclobutrazol 4 ppm drench. Flowering, number of branches, and shoot weight were not different in 'Harvest Moon' treated with height control PGRs compared to controls. Market rating of 'Harvest Moon treated with uniconazole sprays or paclobutrazol drench was higher than controls.

In this study we saw a different response to height control PGRs in different Echinacea cultivars; Harvest Moon was responsive to height control PGRS and was improved with their application while more work will need to be done to find an effective growth control for Marmalade.

BEES IN POLLINATION AND PLANT GROWTH REGULATION

Becky Langer
Bayer Crop Science, Bee Care Center, Durham, North Carolina

The world's population is expected to reach over 9 billion by the year 2050, greatly increasing the demand for food. Finding solutions to increase yields and protect pollinators has never been so critical to both food production and agricultural sustainability. Honey bees, and other pollinators, are responsible for pollinating many of the fruits, nuts and vegetables necessary for a healthy diet and are important contributors to the rich diversity of flowering plants we enjoy in our landscapes. However, these hard-working insects are facing many challenges including pests, diseases and changes to forage and habitat. Meaningful impacts on pollinator health will take all stakeholders working in a concerted effort. This session will review the challenges bees face and highlight the various ways that stakeholders can work together to help pollinators thrive through science, stewardship and solutions.

THE IMPACT OF TEMPERATURE ON THE FLORAL DEVELOPMENT OF BLUEBERRIES (*VACCINIUM CORYMBOSUM* L.)

Sarah A. Mills, Yong-Lak Park, and Nicole L. Waterland
Division of Plant and Soil Sciences, West Virginia University, Morgantown, WV 26506

The United States is the largest producer of blueberries in the world with a production value of $825 million in 2015. Blueberry is an economically important crop that depends on pollinators for fruit set as well as contains a slew of health-promoting phytochemicals. For a successful production of blueberries, healthy development of flowers is essential. One of the regulators of floral development is temperature. Changes in climate have yielded warmer temperatures impacting flower development and activity of pollinators. Studies on flower development have focused mainly on floriculture crops, despite the dependence of fruit production on successful pollination of viable flowers. In this study the effect of temperature on floral development was investigated. Two cultivars of highbush blueberry (Vaccinium corymbosum L.) 'Blueray' and 'Jersey' were grown at five temperatures 18, 20, 24, 28, and 31 °C under natural irradiance in a greenhouse. The number of floral clusters and the rate of development from bud stage through senescence were determined in addition to the measurements of floral longevity and floral bud abortion. Open flowers were first observed on plants grown under 28 °C and the cooler temperature (18 °C) delayed the timing of anthesis. Plants grown under 20 and 24 °C had the largest number of flower clusters, whereas the number of cluster significantly decreased at 28 °C. The duration of floral development decreased in both cultivars as temperature increased. Both cultivars grown at 24 °C reached 'early bloom' stage 3 to 4 days sooner than at 20 °C. Up to 80% of floral clusters were aborted when grown at 31 °C, while 90% and 80% of floral buds survived at 24 and 28 °C, respectively. Generally, a small increment of temperature significantly affected the rate of floral development in a cultivar dependent manner. Temperature change could not only impact on flower development, but also cause asynchronization between flowers and their pollinators. Potential disruption of the plant-pollinator relationship could result in severe reduction of blueberry production.

SEPARATE AND INTERACTIVE EFFECTS OF LIGHT INTENSITY AND TEMPERATURE ON HEALING OF GRAFTED TOMATO SEEDLINGS

Bizhen Hu and Matt Kleinhenz
Department of Horticulture and Crop Science, The Ohio State University, Agricultural Research and Development Center, Wooster, OH, USA

Light and temperature influence the healing of grafted vegetable seedlings but to extents and in ways that are poorly understood. Typical horticultural practice reduces light levels in the healing chamber during the first few days after grafting in part to moderate its temperature. However, we wonder if this approach is best and consistent with the true needs of newly grafted seedlings. For example, if temperature could be controlled reliably, increasing light levels may enhance photosynthesis, thereby providing more energy for graft healing. Already, the use of light-emitting diodes (LEDs) in controlled environments is decoupling light from temperature control, creating the potential to optimize light and temperature levels individually. Regardless, additional information on light-temperature/healing relations will improve understanding of healing and its management. Three studies were completed with this in mind.

Materials and Methods

'Cherokee Purple'/'Maxifort' splice-grafted tomato seedlings were used in all studies conducted in growth chambers. In Study 1, grafted tomato seedlings were healed under four light intensities (5, 50, 150, 300 μmol/m^2/s photosynthetically active radiation) provided solely by LED (60% blue, 20% red and 20% white) at the same temperature around 25 °C day and night. In Study 2, two temperatures (25/20 and 30/25 °C, day/night) by two light intensities (50 and 150 μmol/m^2/s) were tested during healing. In Study 3, three temperatures (15/25, 25/25, and 35/25 °C, day/night) by two light intensities (150 and 300 μmol/m^2/s) were tested during healing.

To evaluate graft healing, survivorship was recorded 10 days after grafting. Vascular reconnection was monitored by dye movement through the graft union 5 days after grafting. Plant regrowth was measured 10 days after grafting including plant height, scion length, stem diameter at the rootstock and scion, stem and leaf fresh and dry weight, leaf area, specific leaf area, and compactness.

Results and Discussion

In Study 1, survival was not different (above 90%) among treatments. Plant regrowth generally had a positive correlation with light intensity (5 to 300 μmol/m^2/s) at around 25 °C. In Study 2, survival was not different (above 90%) among treatments. Plant regrowth was promoted under 150 versus

50 μmol/m²/s, but not affected by temperature alone. Light and temperature interactions were significant in three out of the twelve variables, and 150 μmol/m²/s combined with 30/25 °C tended to achieve the greatest plant regrowth. In Study 3, main and interactive effects of temperature and light intensity were significant. Survival was higher under the moderate temperature at 25/25 °C regardless of light intensities and under 35/25 °C at 150 μmol/m²/s. The moderate temperature at 25/25 °C under 300 μmol/m²/s achieved the largest aboveground dry weight and compactness among the tested combinations.

We conclude that maintaining dim or dark conditions in the healing chamber soon after grafting can be reconsidered, provided temperature control is reliable. The appropriate range of light intensity can be extended to 300 μmol/m²/s, under the range of temperature from 25/20 °C to 30/25 °C (day/night) which grafted tomato seedlings appear to tolerate effectively.

LIGHT AND PLANT GROWTH — STATE OF THE ART AND FUTURE DIRECTIONS

Kevin Folta
Horticultural Sciences Department, University of Florida, Gainesville, FL

Plant productivity and product quality is an exquisite interplay between genetics and environment. We seek to understand both, and merge these areas to affect plant product quality, focusing mainly on small fruits. We utilize genomics-level gene discovery methods that integrate transcriptomics, analytical chemistry and genetics to hone predictions on candidate genes affecting fruit quality. Once identified, we can provide gold-standard functional validation using transgenic plants. When genes are functionally tied to traits, we then can search for genetic variability among existing genetic lines and help inform breeding decisions to produce more nutritious fruit with better flavors. We're also interested in how light can modulate important plant traits like flavor, nutrition, shelf life and general quality. Our laboratory uses the newest technology in LED-based lighting to devise light "recipes" that can control key plant traits. Our goal is to adjust the flavors, nutrition and quality of plant-derived products using these inexpensive, non-chemical treatments.

GROWTH AND MORPHOLOGY OF VEGETABLE SEEDLINGS UNDER DIFFERENT BLUE AND RED PHOTON FLUX RATIOS USING LEDS AS SOLE SOURCE LIGHTING

Ricardo Hernández[1,2], Tomomi Eguchi[1] and Chieri Kubota[1]
[1]School of Plant Sciences, The University of Arizona, Tucson, AZ, USA; Department of
[2]Horticultural Science, North Carolina State University, Raleigh, NC, USA

Vertical-farming (VF) can be more energy-efficient with light emitting diodes (LEDs) with higher efficiencies than fluorescent lamps (FL). Plant responses to blue (B) and red (R) are species specific. The objective was to evaluate B-R photon flux (PF) ratios for the production of vegetable seedlings. Cucumber (cv. Cumlaude) and tomato (cv. Komeett) were grown under 100 μmol·m^{-2}·s^{-1} PPF, 18 h photoperiod, and 25°C air temperature. B (455-nm peak) and R (661-nm peak) LEDs were used as the sole source lighting and the percent B PF (%B) was varied at 0%, 10%, 30%, 50%, 75%, and 100%. For cucumber, hypocotyl decreased with increasing %B up to 75%. Plants in 0%, 10%, and 100%B showed 156%, 65%, and 297%, respectively longer hypocotyl than in FL control. Dry mass and leaf area decreased with increasing %B from 10% to 75%. Plants under FL had 34% and 73% greater dry mass than in 75% and 0%B, respectively and 48%, 30%, 55% and 68% greater leaf area than in 0%, 30%, 50% and 75B%, respectively. For tomato, hypocotyl also decreased with increasing %B up to 75%. Plants in 0%, 10%, 30%, 50% and 100%B showed 110%, 85%, 46%, 42% and 107%, respectively longer hypocotyl than under FL. Dry mass and leaf area increased with increasing %B up to 30-50%, and then decreased from 50% to 100%B. Plants in FL had 53% and 29% greater dry mass than in 0% and 100%B, respectively and 34% and 40% lower dry mass than in 30% and 50%B. Tomatoes under FL had 57% greater leaf area than 0%B. For cucumber seedlings, the optimal %B was 10%, which, however, produced taller seedlings with less dry mass than those under conventional FL. For tomato, the optimal %B was 30-50%, which produced seedlings with comparable morphology and growth as those under FL.

LED LIGHTING UNDER VARIOUS SOLE-SOURCE AND SUPPLEMENTAL SCENARIOS AFFECTS MINI-ROSE ELONGATION AND FLOWER GROWTH

Scott Malain[1], Lola Siboldi[2] and Melanie Yelton[1]
[1]LumiGrow, Emeryville, CA,
[2]Rocket Farms, Half Moon Bay

Experiments using LEDs to apply various wavelengths of light under greenhouse and sole-source scenarios report inconsistent effects on elongation and growth of plants. Exploring the growth of plants under different spectrum may allow for a better understanding of wavelengths effect.

Varietal differences often complicate elongation control strategies due to variation in morphology. Currently, plant growth regulator chemicals are used in various quantities, depending on the variety, to achieve the designated height desired. It follows that the use of various wavelengths of light will also exhibit varietal differences in final height. For mini-roses, the varieties 'Karina' and 'Renata' typically express large differences elongation and flower production, with 'Karina' expressing a shorter stature with a higher flower count when compared to 'Renata.' Growing these two varieties of roses under the experimental lighting regimes will provide useful data for varietal differences.

'Karina' and 'Renata' mini-roses were grown under three spectrum provided by LumiGrow Pro 325 LED fixtures with three different lighting scenarios. The three lighting scenarios were sole-source LED light, artificial solar (provided by MH) with supplemental LED, and in a greenhouse with supplemental LED light. In all scenarios, total DLI, including incoming light, was conserved at 10.8 $\mu mol \cdot m^{-2} \cdot s^{-1}$. The light provided from the LEDs consisted of three ratios of red:white:blue, and were 9:1:0, 7:1:2, and 5:1:4.

As expected, 'Renata' was consistently taller than 'Karina' in every treatment. Under the sole-source LED treatments, a consistent reduction in elongation was observed in both varieties with increasing blue light. The 9:1:0 sole-source 'Karina' exhibited the most elongation while the 5:1:4 sole-source plants grew shortest across any other lighting treatment. Under the artificial solar and greenhouse treatments, plants did not exhibit the same elongation patterns with increasing blue in comparison to the sole-source treatments.

In order to better understand light spectra in any commercial setting, an experiment to measure the wavelength of light under different greenhouse glazing materials is in a planning phase.

Small greenhouses are being constructed with the most prevalent greenhouse glazing materials. Light spectra will be measured on set intervals in the greenhouses and outdoors throughout the year at several locations across the U.S. and Canada to build an understanding of light spectra in growing environments. These data will then be used to correlate observed plant morphologies in these settings to understand light effects.

PLANT GROWTH REGULATION AND NUTRIENT INTERACTION

Daniel I. Leskovar, Yahia Othman and Vijay Joshi
Texas A&M AgriLife Research, Texas A&M University, Uvalde, TX 78801

Plant growth regulators (PGRs) modify plant morphology through changes in cell division and cell expansion, and ultimately the vegetative and reproductive development. We found positive effects/impacts on growth, water relations and physiological conditioning of foliar ABA application in pepper, cantaloupe and artichoke transplants. ABA was also an effective growth holding agent, extending the marketability of pepper in the nursery. Studies in Spain showed GA improving plant height and quality, while auxin alone or combination with GA and CK increased yield in sweet pepper. Other research emphasized the importance of auxin and GA interaction in fruit set. PGRs involve complex nutritional processes e.g. by nitrate-N and other hormonal signaling pathways that are linked at several levels of integration. This review will first highlight key root/shoot responses to single or multiple PGRs in vegetable crops, and then provide examples of the interactive effects of PGRs and nitrogen in model systems and commercial crops. Our current interest is in the application of bioregulators that can promote root/shoot development and enhance water and heat stress tolerance.

THE EFFECTS OF ABSCISIC ACID ON MINERAL NUTRIENT UPTAKE IN TOMATO LEAF AND FRUIT TISSUE

T. Casey Barickman[1], Dean A. Kopsell[2], and Carl E. Sams[2]
[1]Mississippi State University, North Mississippi Research and Extension Center, Verona, MS 38879.
[2]The University of Tennessee, Department of Plant Sciences, Knoxville, TN 37996

Plants encounter various environmental stress factors that can impact nutritional requirements and fruit quality. Adequate levels of mineral nutrients in tomato (Solanum lycopersicum) fruit have positive effects on fruit quality. One of the results of insufficient mineral nutrient uptake and movement in tomato fruit is the physiological disorder blossom-end rot (BER), which is associated with a calcium (Ca) deficiency in the distal fruit tissue. ABA effects gas exchange by closing stomates, therefore decreasing photosynthesis which has a negative effect on vegetative growth. However, ABA can have a positive effect on nutritional fluxes in the plant. For example, ABA can enhance potassium (K) absorption in cucumber (Cucumis sativus) under high temperature conditions (Du and Tachibana, 1995), Previous research has also demonstrated that foliar abscisic acid (ABA) applications increased the uptake of Ca into fruit tissue and decreased the incidence of BER (Barickman et al., 2014). Furthermore, Perez-Jimenez et al. (2015) found that applications of ABA increased iron (Fe) in sweet pepper (Capsicum annuum) fruit tissue. The role of ABA has been demonstrated with Fe storage and remobilization in cucumber (Cucumis sativus) leaf tissue (Bacaicoa et al., 2009). Other research has indicated that ABA does not have a role in the regulation of Fe uptake in Arabidopsis thaliana (Schmidt et al., 2000). This study examined how root and foliar spray ABA applications, individually and in combination, affect the partitioning of mineral nutrients between the leaves and fruit of tomato plants, especially in the distal tissue, and how ABA affects the incidence of BER in the distal tissue of tomato fruit. 'Mt. Fresh Plus' tomato plants were grown in the greenhouse at 25/20 °C (day/night) under a 16 h photoperiod. Plants were treated with different Ca concentrations in the fertilizer solution. Plants were also treated with foliar spray ABA applications weekly. Calcium was applied via the irrigation lines at 60, 90, or 180 mg·L^{-1}. ABA treatments were applied as a combination of foliar sprays and root applications. Foliar ABA application treatments consisted of de-ionized (DI) water control (0.0 mg ABA·L^{-1}) or 500 mg ABA·L^{-1}. For ABA root applications, treatments consisted of a DI water control (0.0 mg ABA·L^{-1}) or 50 mg ABA·L^{-1} applied via the irrigation lines. ABA spray treatments were applied once weekly until dripping from the foliage, while root applications were applied four times per day through the irrigation system. Fruit tissues were harvested 84-90 d after seeding. Fruit tissue was harvested at red ripe maturity and evaluated for yield, BER, and mineral nutrient concentrations. Leaves were harvested at time of fruit and were analyzed for mineral nutrient concentrations. The results indicate that a combination of the spray and root applications of ABA resulted in the greatest decrease in

BER. The foliar spray application of ABA combined with the Ca treatment of 180 mg·L^{-1} decreased the incidence of BER. Results also demonstrate that ABA treatments are effective in increasing fruit Ca and preventing BER in the early stages of plant development, but are less effective in preventing Ca deficiency in the later stages of growth. In tomato leaf tissue, applications of ABA generally decreased mineral nutrients, such as K, boron (B), and zinc (Zn). However, root applications of ABA increased phosphorus (P) and molybdenum (Mo). In the fruit tissue, ABA treatments increase B, manganese (Mn), Mo, P, and K concentrations. For example, root ABA applications increased Mo and B concentrations by 25% when compared to the control treatment. In addition, foliar ABA applications increased Mn 16% when compared to the control treatment. The partitioning of mineral nutrients in the proximal and distal fruit tissue were significantly different. For example, B, Fe, Mo, magnesium (Mg), and P concentrations were higher in the distal fruit tissue. Overall, applications of ABA affected the mineral nutrient in tomato leaf and fruit tissues. Thus, ABA could be a potential novel treatment to impact mineral nutrient uptake, partitioning, and distribution in tomato leaf and fruit tissue.

Literature Cited:

Barickman, T.C., D.A. Kopsell, and C.E. Sams. 2014. Foliar applications of abscisic acid decrease the incidence of blossom-end rot in tomato fruit. Scientia Hort.

Du, Y.C. and S. Tachibana. 1995. Effects of supraoptimal root temperature on ABA levels in cucumber plants and its control by ABA applied to roots. Acta Hort. 394: 227-234.

Perez-Jimenez, M. M. Pazos-Navarro, J. Lopez-Marin, A. Galvez, P. Varo, and F.M. del Amor. 2015. Foliar application of plant growth regulators change the nutrient composition of sweet peppers (Capsicum annum L.). Sci. Hort. 194:188-193.

Schmidt, W., J. Tittel, and A. Schikora. 2000. Role of hormones in the induction of iron deficiency response in Arabidopsis thaliana roots. Plant Physiol. 122: 1109-1118.

EFFECT OF UREA AND A COMMERCIAL EXTRACT OF ASCOPHYLLUM NODOSUM ON BUD ABSCISSION IN PISTACHIO.

Holly Little[1] and Gurreet Brar[2]
[1]Acadian Seaplants
[2]University of Florida

Pistachios have a unique mechanism of alternate bearing, where excessive flower bud abscission occurs during the “on year” (heavy crop). Bud abscission begins during the initiation of embryo development of the current crop (June) and continues through embryo growth. The physiological development of the alternated bearing phenomenon in pistachio is not fully understood, although there is evidence that nutrients, carbohydrates, and plant growth regulators all influence it. Alternate bearing is a big challenge for pistachio growers. Previous work has shown that nitrogen, cytokinins, and seaweed extract can positively influence bud abscission in pistachio. In 2014, a multi-year trial was initiated to test a program, that was developed by a grower, to reduce bud abscission in pistachio. Treatments included a water control, Acadian LSC (Acadian Seaplants LTD) at 9.35 L/ha, low biuret urea at 44.83 kg/ha (20.6 lb N/ha), and the combination of Acadian LSC and urea. All treatments were applied in 2806 l of water/ha. Two applications of each treatment were made each year; end of June and end of July in 2014, and end of May and end of June in 2015. In 2014, there was little bud abscission, and all treatments increased bud formation, with the highest increase in the Acadian LSC treatment, followed by the urea. In 2015, there was approximately 35% bud abscission in the control, and the Acadian LSC had the lowest percent bud abscission (20%), followed by the urea (28%). The combination of the Acadian LSC and urea had abscission levels similar to the control, and visual symptoms of phytotoxicity were seen. In both years the Acadian LSC alone wat the only treatment to be statistically different from the control. The rate of urea in this study was much higher than in the previous work, which may have contributed to it being less effective.

BOOSTING PLANT DEFENSES BY BENEFICIAL SOIL FUNGI

María J. Pozo, Javier Rivero, Ainhoa Martinez Medina, Sabine Jung, Ivan Fernández, Judith Pozo, Juan M. García, Juan Antonio López Ráez, Concepción Azcón-Aguilar, José-Miguel Barea
Department of Soil Microbiology and Symbiotic Systems. Estación Experimental del Zaidín (CSIC), Profesor Albareda 1, 18008 Granada, Spain.
E-mail address: mariajose.pozo@eez.csic.es

For survival, plants have to efficiently adjust their phenotype to environmental challenges, and they do that by finely coordinating their responses to balance their growth and defence performance. Such phenotypic plasticity largely relies on phytohormone signaling, that can be modulated by the plant associated microbiota. Beneficial microbes are common in the root-associated soil, and their interaction with the plant usually promotes plant growth and increase the plant´s ability to cope with biotic and abiotic stress factors. Among these beneficial microorganisms, arbuscular mycorrhizal (AM) fungi are of major importance as they establish mutualistic symbiosis with most plant species. The widespread mycorrhizal symbioses modify plant responses to external stimuli, generally improving the resilience of the symbiotic system to environmental constraints. Protective effects of the AM symbiosis against pathogens, pests, and parasitic plants have been described for many plant species, including agriculturally important crop varieties. Besides mechanisms such as an improved plant nutrition and competition, experimental evidence supports a major role for the priming of plant defence mechanisms in the evidenced protection. Indeed, during mycorrhiza establishment plant defence mechanisms have to be tightly regulated in order to achieve a functional symbiosis, and host hormone homeostasis is altered. As a consequence of this regulation, a mild, but effective activation of the plant immune system occur, not only locally but also systemically. This activation seems to underlie the "alert" or "primed"state of the plant that allows for a more efficient activation of defence mechanisms in response to a potential attack.

Introduction

Microbial populations living at the root-soil interfaces -the rhizosphere- are known to influence plant health, and soil quality, key issues for agro-ecosystem sustainability (Barea et al. 2013a). They are able to affect plant growth and to influence plant responses to the environment. Beneficial microbes can enhance plant fitness by alleviating the effect of stress factors. They can directly remove or diminish the stress, for example by increasing nutrient availability or antagonizing pathogens, but they can also affect plant phenotypic plasticity in a broad range of traits towards improved resistance or tolerance to stress situations. Generally, these effects are at least partly mediated by changes in plant hormone homeostasis. Interestingly, plants can also regulate their interaction with microbes according to external conditions, usually also through phytohormone regulated mechanisms, promoting some over others, and even modulating microbial behavior (Fig. 1) (Pozo et al. 2015).

Like pathogens, beneficial microorganisms are confronted with the innate immune system of the roots and colonization success essentially depends on the evolution of strategies for immune evasion. The modulation of plant defence responses by microbial symbionts aids establishing the delicate balance between the two partners, and may result in an enhanced defensive capacity of the plant (Zamioudis and Pieterse 2012, Selosse et al., 2014). This enhanced resistance is not only restricted to the roots, but it is also expressed in aboveground plant tissues, thereby providing the plant with an induced systemic resistance effective against a broad spectrum of pathogens (Pieterse et al., 2014)

The AM symbioses can be found in almost all agro- and eco-systems worldwide to improve plant growth and health through key ecological processes (Barea et al. 2013b). The AM symbiosis increases plant growth, enhances available soil nutrient uptake, increases plant resistance to drought and protect plants against a wide range of below-ground attackers such as soil-borne fungal and bacterial pathogens, nematodes or root-chewing insects and induced resistance against shoot pathogens. (Barea et al., 2013b; Jung et al. 2012). Therefore, the AM fungi, a major group of plant mutualistic endophytic microorganisms, can be considered as "root allies", which support plants to cope with biotic stressors. The AM symbiosis can be managed, as a low-input biotechnology, to help sustainable environmentally-friendly agro-technological practices.

Protection against soil borne pathogens

The AM symbioses is able to reduce the damage caused in crops by many soil borne pathogens, including major pathogens from the genera Fusarium, Rhizoctonia, Verticillium, Phytophthora and Pythium, parasitic nematodes and root feeding insects (Koricheva et al. 2009; Jung et al. 2012).

Several mechanisms operate simultaneously in the alleviation of the damage caused by soil borne pathogens (Jung et al., 2012). For example, AM fungi have been proven to directly compete with soil borne pathogens and nematodes for space and nutrients and to alter root morphology, what may change the dynamics of pathogen infection. Mycorrhizal colonization also alters the quality and quantity of root exudates impacting the microbial community of the mycorrhizosphere. These changes may lead to a shift in its composition favouring microbial biocontrol agents (Barea et al. 2013a). Moreover, in well-established mycorrhizas the secretion of strigolactones –germination stimulants for parasitic weeds- to the rhizosphere by the host roots is reduced, so this change in root exudation reduce the incidence of parasitic plants as Orobanche (López Raez et al. 2012).

The use of experimental split-root systems, allowing physical separation of the beneficial microbe and the aggressor, has confirmed that the protection by mycorrhiza is manifested also in non-colonized areas of the root system, and consequently, in addition to competition, the protection is mediated by plant defense mechanisms (Pozo et al. 2002; Martinez Medina et al., 2016b).

Systemic protection against leaf attackers

These plant mediated effects of the mycorrhiza affect not only to roots, but also to shoot interactions, although the impact depends on the attacker lifestyle or feeding guild. The symbiosis is mostly efficient in enhancing resistance to necrotrophic shoot pathogens as Botrytis cinerea and Alternaria spp. This protection is related to the stronger activation of jasmonic acid (JA) regulated plant defense responses, as it occurs during interactions with other beneficial microbes (Fig.2) (Jung et al., 2012; Martinez Medina et al., 2013; Martinez-Hidalgo et al. 2015, Sanchez Bel et al., 2016).

In relation to insects, the outcome of the AM-plant-herbivore interaction depends on many factors, such as the AM fungus, host plant, insect species involved and prevailing environmental factors (Pineda et al. 2010). The AM symbiosis can actually influence insect herbivore performance, but the magnitude and direction of the effect depend mainly upon the feeding mode and life style of the insect (Koricheva et al. 2009). The improvement of plant nutrition and of plant defence by the symbiosis may have opposite effects: on one hand by improving nutrient and water uptake, mycorrhizas can facilitate the re-growth of tissues after herbivory by promoting plant tolerance through compensation of biomass losses. On the other hand, as nutrition improves, plants may become more nutritive or attractive to insects. Furthermore, the induction of defence mechanisms operating in resistance against microbial pathogens may also impact herbivorous insects (Jung et al. 2012). The final impact on insect performance will depend on the interplay between a positive effect derived from the enhanced plant quality and a negative effect derived from the induced resistance in the plant. In general terms, mycorrhizal plants are more resistant to generalist chewing insects, as these insects feed on the leaf tissue and cause massive damage which activates defences that depend on the plant hormone JA (Fig.2) (Jung et al., 2012, Song et al., 2013)

Modulation of the plant's immune system and induced systemic resistance

Both mutualistic and pathogenic endophytes are initially recognized as alien organisms and the plant reacts with the activation of an immune response. Indeed, AM fungi have to deal with the plant's immune system for root colonization. Once the symbiosis is established, the plant has to regulate the level of AM fungal proliferation within the roots according to different environmental conditions as nutrient and light availability to prevent excessive colonization and carbon drainage. The plant control over mycorrhizal colonization is mediated by changes in hormone homeostasis (Fig. 1) (Fernandez et al., 2014, Pozo et al., 2015). Changes in hormone levels and in the profile in multiple primary and secondary metabolites related to defence have been reported in mycorrhizal roots (Rivero et al., 2015). This modulation of plant defence signalling pathways may result in pre-conditioning of the tissues for efficient activation of plant defences upon a challenger attack, a phenomenon that is called priming (Fig.3) (Conrath et al., 2006; Pozo and Azcón-Aguilar 2007; Martínez-Medina et al., 2016a).

Indeed, regulation of JA has been reported to have a central role in the correct functioning of the AM symbiosis (Hause and Schaarschmidt 2009). The regulation of this defence signalling pathways may lead to the modulation of plant immunity by AM fungi and play a major role in Mycorrhiza Induced Resistance (MIR). As mentioned, this induced defence is also expressed in aboveground plant tissues, giving rise to a systemic response that is typically effective against a broad spectrum of plant pathogens and even herbivores. The dependence of successful mycorrhization on the control of JA and salicylic acid (SA) signalling would explain the range of protection conferred by this symbiosis (Pozo and Azcón-Aguilar 2007, Jung et al. 2012). Mycorrhizal plants are, in general, more resistant to necrotrophs and chewing insects, which are targeted by JA-dependent defence responses.

In summary, this priming of the plant's innate immune system is common upon interaction with beneficial microorganisms and has important fitness benefits compared to direct activation of defences. Priming set the plant is an "alert" state in which defences are not actively expressed but in which the response to an attack occurs faster and/or stronger compared to plants not previously exposed to the priming stimulus, efficiently increasing plant resistance. Thus, priming confers important plant fitness benefits (Fig. 3) (Conrath et al. 2006, Martínez-Medina et al., 2016). A primed response of JA-dependent defences was confirmed by transcript profiling of leaves and the use of tomato mutants impaired in JA signalling confirmed that JA is required for induced resistance against Botrytis by AM and other beneficials as Trichoderma spp, and plant growth promoting rhizobacteria such as Micromonopora spp and Pseudomonas spp. (Martínez-Medina et al., 2013, Martínez-Hidalgo et al., 2015 Pieterse et al., 2014, Sanchez-Bel et al., 2016), corroborating that the "induced resistance" by beneficials requires a functional JA signalling pathway for the efficient induction of resistance

Conclusions and further research

Beneficial soil microbes, and particularly AM fungi, are key element in natural and man-made ecosystems. The establishment of the AM symbiosis with plant roots significantly alters the host plant physiology and has far reaching consequences for the plant and its biotic interactions. Generally, mycorrhizal symbioses enhance the plant´s ability to cope with abiotic and biotic stresses. Even though the individual outcome always depends on the AM fungal-plant attacker combination, protective effects against deleterious organisms ranging from microbial pathogens to herbivorous insects and parasitic plants have been widely described.

Experimental evidence confirms that this protection is based not only on an improved nutrition or local changes within the roots and the rhizosphere, but also on an efficient priming of plant immunity. In fact, mycorrhizal colonization boosts the plant ability to respond to attackers by priming JA-dependent defenses. The identification of defense regulatory elements coordinating mycorrhizal development and MIR is a major challenge for research. This identification is paving the way to the development of biotechnological strategies for improving mycorrhiza establishment and the use of AM fungi in the integrated management of pests and diseases.

References

Azcón-Aguilar, C. & Barea, J. Arbuscular mycorrhizas and biological control of soil-borne plant pathogens – an overview of the mechanisms involved. Mycorrhiza (1997) 6: 457-464.

Barea J. M., Pozo M. J., Azcón R., Azcón-Aguilar C. (2013a). Microbial Interactions in the Rhizosphere. In: Molecular Microbial Ecology of the Rhizosphere (de Bruijn, F. Ed.) Wiley-Blackwell. pp: 29-44.

Barea, J.M., Pozo, M. J., López-Ráez, J. M., Aroca, R., Ruíz-Lozano, J. M., Ferrol, N., Azcón, R., Azcón-Aguilar, C. 2013b. Arbuscular Mycorrhizas and their significance in promoting soil-plant systems sustainability against environmental stresses. In: Beneficial Plant-Microbial Interactions: Ecology and Applications (Rodelas, B. and Gonzalez-Lopez, J., Eds.). CRC Press, USA. Chapter 15, pp. 353-387.

Conrath U. et al., (2006) Priming: getting ready for battle. Molecular Plant-Microbe Interaction, 19: 1062-1071

Fernández I., Merlos M., López-Ráez J.A., Martínez-Medina A., Ferrol N., Azcón C., Bonfante P., Flors V., Pozo M.J. Defense related phytohormones regulation in arbuscular mycorrhizal symbiosis depends on the partner genotypes (2014). Journal of Chemical Ecology 40 (7): 791-803

Jung S, Martínez-Medina A, López-Ráez JA, Pozo MJ (2012) Mycorrhiza-induced resistance and priming of plant defenses (2012). Journal of Chemical Ecology 38(6): 651-664.

López-Ráez J. A., Pozo M. J., García-Garrido J. M. (2011) Strigolactones: a cry for help in the rhizosphere (2011). Botany 89(8): 513-522.

Martínez-Hidalgo P, García JM, Pozo MJ (2015) Induced systemic resistance against Botrytis cinérea by Micromonospora strains isolated from root nodules. Frontiers in Microbiology. Volume 6, Issue SEP, Article number 922

Martinez-Medina, A., I. Fernandez, Sánchez-Guzmán, M.J, Jung, S., Pascual, J. A, Pozo, M. J. (2013). Deciphering the hormonal signalling network behind the systemic resistance induced by Trichoderma harzianumin tomato. Frontiers in Plant Science 4: 206.

Martínez-Medina A, Flors V, Heil M, Mauch-Mani B, Pieterse CM, Pozo MJ, Ton J, van Dam NM, Conrath U (2016a). Recognizing plant defense priming. Trends in Plant Science 21(10): 818-822

Martínez-Medina A, I Fernández I, Lok GB, Pozo MJ, Pieterse CMJ, Van Wees SM (2016b) Shifting from priming of salicylic acid- to jasmonic acid-regulated defences by Trichoderma protects tomato against the root knot nematode Meloidogyne incognita. New Phytologist.

Pieterse, C. M. J., C. Zamioudis, et al. (2014). Induced systemic resistance by beneficial microbes. Annual Review of Phytopathology. 52: 347-375.

Pineda, A., Dicke, M., Pieterse C.M.J., Pozo M.J. (2013) Beneficial microbes in a changing environment: are they always helping plants to deal with insects? Functional Ecology 27(3): 574-586.

Pozo M. J., Cordier C., Dumas-Gaudot E., Gianinazzi S., Barea J. M., Azcón-Aguilar C. (2002). Localized vs systemic effect of arbuscular mycorrhizal fungi on defence responses to Phytophthora infection in tomato plants. Journal of Experimental Botany, 53: 525-534

Pozo MJ, López-Ráez JA, Azcón-Aguilar C, García-Garrido JM (2015). Phytohormones as integrators of environmental signals in the regulation of mycorrhizal symbioses. New Phytologist 205: 1431-1436

Rivero J, Gamir J, Aroca R, Pozo MJ, Flors V (2015). Metabolomic transition in mycorrhizal tomato roots. Frontiers in Microbiology 6 (JUN), Article number 598

Sanchez-Bel P, Troncho P, Gamir J, Pozo MJ, Camañes G, Cerezo M, Flors V. (2016). The nitrogen availability interferes with mycorrhiza-induced resistance against Botrytis cinerea in tomato. Frontiers in Microbiology 7(1598)

Selosse MA, Bessis A, Pozo MJ (2014). Microbial priming of plant and animal immunity: symbionts as developmental signals. Trends in Microbiology 22(11): 607-613

Song, Y. Y., M. Ye, et al. (2013) Priming of anti-herbivore defense in tomato by arbuscular mycorrhizal fungus and involvement of the jasmonate pathway. Journal of Chemical Ecology 39(7): 1036-1044.

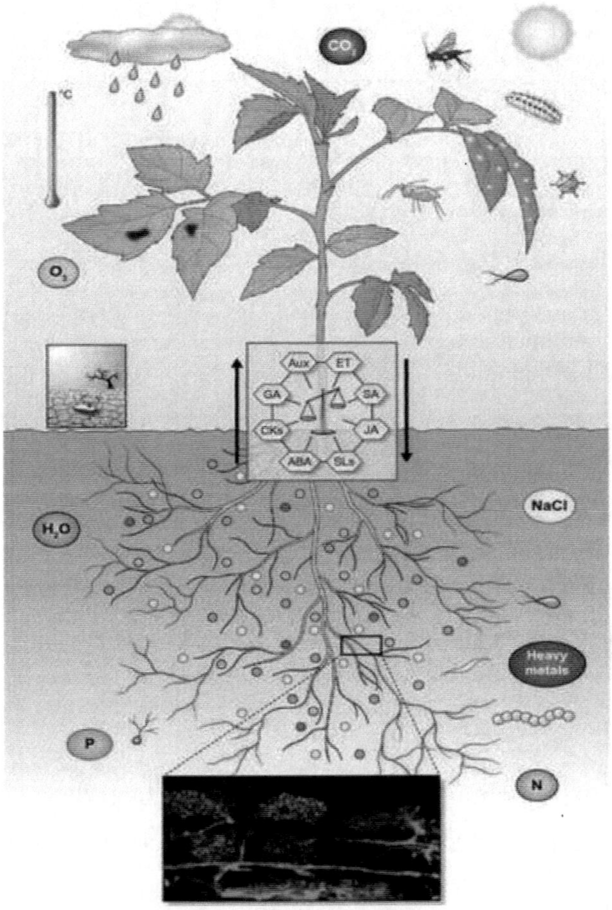

Figure1. Phytohormone networks integrate plant responses to the different environmental cues and mycorrhiza formation and function. They mediate the plant control on mycorrhizal colonization according to the environmental conditions such as nutrient and light availability. On the other hand, hormone homeostasis is modulated in mycorrhizal plants, altering the plant response to stresses and biotic interactions.

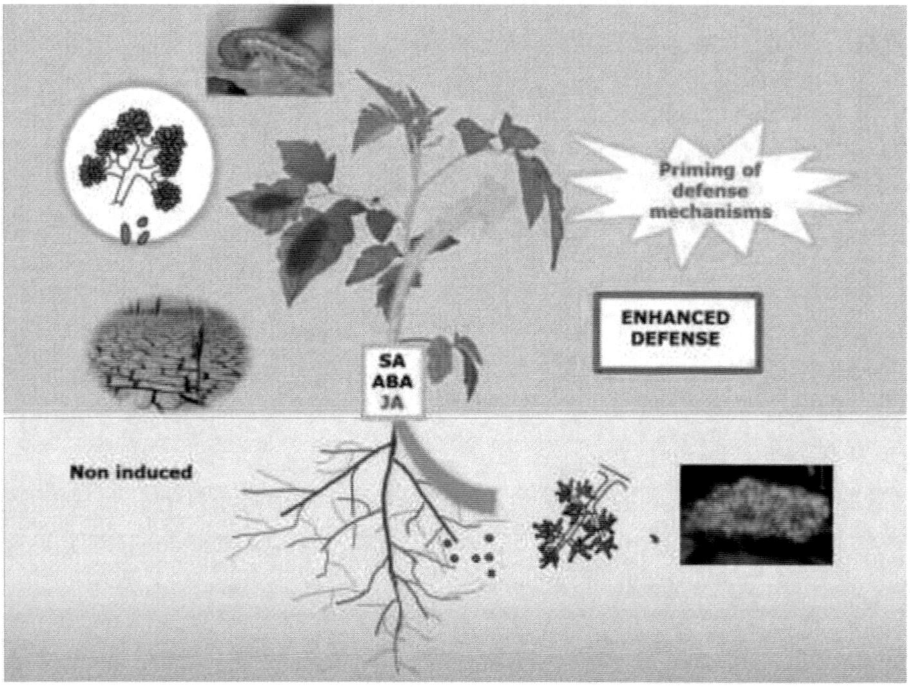

Figure 2. Model for the induction of resistance against shoot atackers and biotic stress tolerance by priming jasmonate dependent defense related signaling. Root colonization with beneficial fungi such as Trichoderma and mycorrhizal fungi primes leaf tissues for enhanced activation of JA regulated defenses leading to a high resistance to necrotrophic pathogens and chewing insects. Frequently, the functionality of other signaling pathways is required for the priming of JA defenses.

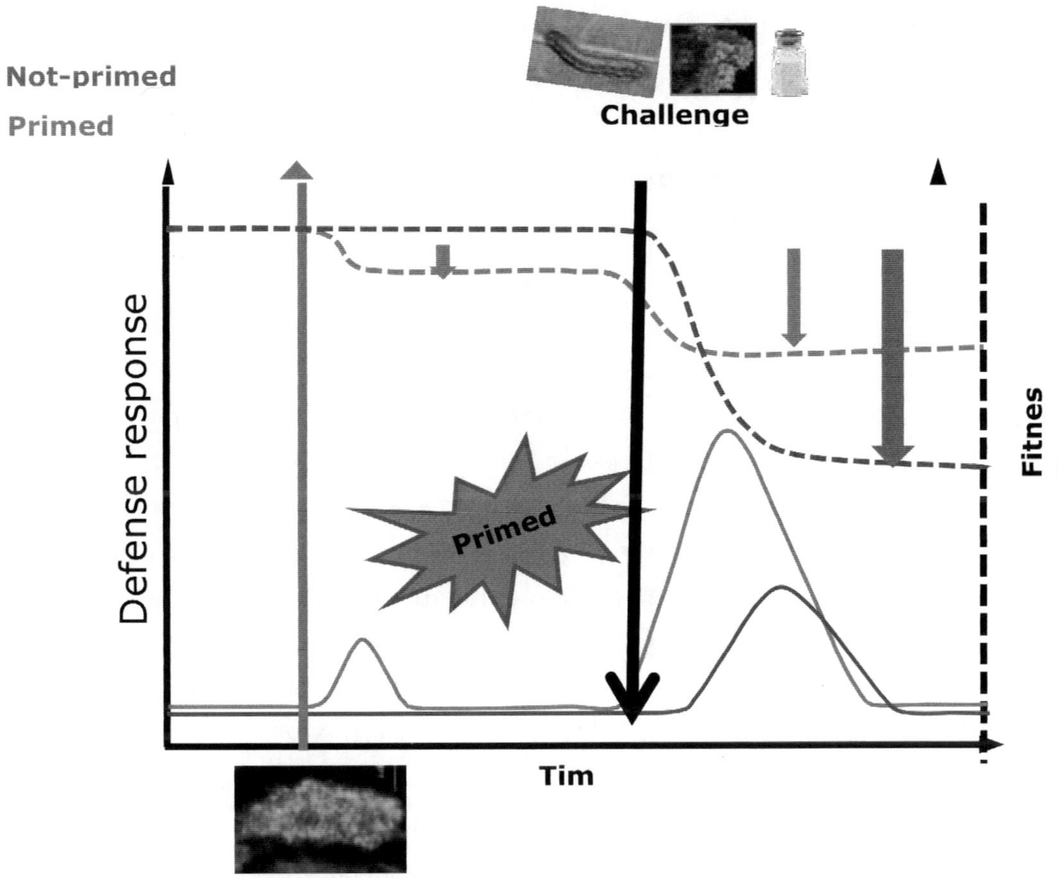

Figure 3. Defense intensity and plant fitness in priming of defenses by beneficial microbes. Beneficial microbes may trigger a small and transient activation of plant defense mechanisms that results in an "alert" or "primed" state of the host. The maintenance of the primed state (before the challenging stress) is expected to have low fitness costs as compared to the direct activation of defenses. Frequently, the growth promoting effects of the beneficial microbe compensate this cost.In response to the challenge (biotic or abiotic stress factors), primed plants mobilize defenses faster, earlier and/or stronger than unprimed plants. As a consequence, the impact of the stressor on plant fitness is significantly reduced. Defense responses (solid lines) and fitness (dashed lines) in primed (red) vs. unprimed (blue) plants. Modified from Martinez-Medina et al., 2016.

PROHEXADIONE-CALCIUM RESPONSIVE ALFALFA VARIETIES ENSURE SUCCESS OF CORN-INTERSEEDED ALFALFA PRODUCTION SYSTEMS

John H. Grabber[1], Mark J. Renz[2], and Heathcliffe Riday[1]
[1]U.S. Dairy Forage Research Center, USDA-ARS, Madison, WI 53706
[2]Agronomy Department, University of Wisconsin, Madison, WI 53706

Recent USDA-NASS data indicate alfalfa and silage corn were planted on about 0.8 and 1.9 million hectares per year, respectively, in the Northeast, Great Lakes, Upper Midwest, and Northern Mountain regions of the USA. Because both crops are often grown in rotation, alfalfa could be interseeded at corn planting throughout these regions to serve as a dual purpose crop for providing groundcover during silage corn production and forage during subsequent growing seasons. Unfortunately, this system has been unworkable because competition between the co-planted crops often leads to stand failure of interseeded alfalfa. Recent Wisconsin studies demonstrated that properly timed foliar applications of prohexadione-calcium (PHD) doubled plant survival of alfalfa during establishment in densely planted silage corn. The following year, the average dry matter yield of interseeded alfalfa treated with PHD was two-fold greater than conventionally spring-seeded alfalfa. In other work, interseeded alfalfa reduced fall and spring runoff of water and phosphorus by 60% and soil erosion by 80% compared to cropland containing only silage corn residues and weeds. These improvements in crop yields and in soil and water conservation are powerful incentives for implementing this production system on farms, but poor seedling survival of interseeded alfalfa at some locations, even with PHD treatment, proved to be a major barrier blocking implementation of this system. Fortunately, a major breakthrough occurred in 2015 when it was discovered that conventional and glyphosate-resistant alfalfa varieties differ substantially in both plant survival and response to PHD when interseeded into corn. In this study, plant density of 20 alfalfa varieties in October, four weeks after corn harvest, ranged from 18 to 90 plants m-2 without PHD and from 55 to 210 plants m-2 with PHD treatment. Plant density responses of varieties to PHD treatment ranged from not significant to an increase of nearly 4-fold. Plant survival of alfalfa was not associated with plant characteristics measured during the growing season or with most plant traits. However, two alfalfa varieties with the leafhopper-resistance trait ranked first in plant survival. Overall, these results suggest the corn-interseeded alfalfa production system could be implemented over a wide array of locations if an appropriate alfalfa variety is sown and treated with PHD.

REGULATION OF FLORAL STEM CELL IN ARABIDOPSIS: SIGNALING PLAYERS AND THE CROSS-TALKS WITH AUXIN

Zhigang Huang, Wanhuang Lin, Ruozhong wang and Langtao Xiao
Hunan Provincial Key Laboratory of Phytohormones and Growth Development, College of Bioscience and Biotechnology, Hunan Agricultural University, Changsha, China

Floral stem cells reside in floral meristems and give rise to all organs in a flower. In Arabidopsis and many other plants, the fate of floral stem cells is precisely regulated such that stem cell maintenance is terminated upon the production of a full complement of organ primordia. The transcription factor AGAMOUS (AG) directly represses the stem cell maintenance gene WUSCHEL (WUS). AG also activates the expression of KNUCKLES (KNU), which in turn represses WUS expression. Thus, AG, KNU and WUS form a feed-forward module to terminate floral stem cell maintenance. We performed a genetic screen in the ag-11 background and isolated mutations that suppressed the floral determinacy defect. Two were found to harbor mutations in AP2, and the results showed that AP2 antagonizes AG in terms of WUS repression in the floral meristem. Auxin also plays an essential role in specifying the identity of floral organs. AUXIN RESPONSE FACTOR 3 (ARF3) is a direct target of AP2 and partially mediates AP2's function in FM determinacy. The DR5::GFP and DII-VENUS reporter lines will be integrated into the floral determinacy mutants. Further analysis of the relationship between the auxin pathways and the floral stem cell control genes will provide critical information regarding mechanisms of floral meristem determinacy.

Supported by National Natural Science Foundation of China, Grant 31570372

SOYBEAN (*GLYCINE MAX*) GROWTH AND YIELD RESPONSES TO EARLY SEASON APPLICATION OF GIBBERELLIC ACID

Michael D. Rethwisch, Craig J. Hruska, Scott Willet and Garrett Tooker
University of Nebraska-Lincoln, Butler County Extension, 451 N. 5th Street, David City, Nebraska, USA 68632-1666
Email: mrethwisch2@unl.edu

Gibberellic acid (GA3) was applied to soybeans (Glycine max) at unifoliate (V-U), first (V-1), second trifoliate leaf (V-2) or sequentially to two early stages of development to determine the potential benefit of this chemistry. Plant growth data were collected at regular intervals throughout the growing season. Internode lengths and percent branching of unifoliate nodes were noted to be significantly affected with unifoliate and first trifoliate leaf sequential applications, but percent branching was not increased with applications initiated at first trifoliate leaf. Increases in unifoliate node branching contributed to significantly more pods/plant at harvest (mean of 22.9/plant, $P<0.001$). Differences in pods/plant were not noted from single applications treatments. Yields due to sequential applications initiated V-U were not significantly different and did not correspond with pod counts noted in August, although more pods numbers at harvest were still documented. Yields were affected by soil environment/ fertility levels, with sequential applications of gibberellic acid resulting in reduced yield at a low fertility site, while a large increase (12%) was noted in a high yield environment. A single application of gibberellic acid at first trifoliate leaf resulted in a highly significant yield increase (17%, $P<0.0001$), although pod numbers/plant were similar. Mean yield increase of 430 kg/ha (6.4 bushels/acre) with a mean net value increase of $37.40/acre was noted for single applications at V-1 and V-2.

Introduction

Previous local small plot field research had documented significant increases in soybean internode lengths from single applications along with the surprising discovery of increased pods/plant at harvest for two (2) sequential applications of gibberellic acid (GA-3). While pod data were collected, yields were not obtained in initial small plot studies.

As there are no known scientific data available/articles published on soybean growth and yield responses to application of GA-3 form of gibberellic acid at early stages of development, the need exists for replicated experiments to document effects of soybean growth and yield response to early season application(s) of GA-3.

Methods

A series of field experiments were conducted during 2014-2015 to document soybean responses to early season application of GA-3. These consisted of one (1) multi-site small plot experiment, two (2) large plot 2014 experiments and two (2) large plot 2015 experiments, each with three or four replications at each site. GA-3 formulation used in all experiments was RyzUp SmartGrass® 40 WDG (Valent, USA).

A) RyzUp SmartGrass® applied at 28 g/ha (4 oz./acre) at unifoliate leaf (V-U), and 21 g/ha (3 oz./acre) seven days later at first trifoliate leaf (V-1). This experiment was conducted at four (4) locations utilizing small plots (4 rows wide (3.1 wide x 12.2 meters long). Variety used at each site was Asgrow 2733 (2014)
B) RyzUp SmartGrass® applied at 21 g/ha (0.3 oz./acre) at unifoliate leaf (V-U), and 21 g/ha four days later at first trifoliate leaf (V-1). Large plots (27.4 meters wide x field length). The variety at this site (Ulysses) was Syngenta 24K2, and was combine harvested. (2014)
C) RyzUp SmartGrass® applied at 21 g/ha (0.3 oz./acre) at V-1 and 21 g/ha four days later at second trifoliate leaf (V-2). The variety at this site (Rising City) was NK S25-E4. Plots were 24 rows wide x field length and combine harvested (2014)
D) RyzUp SmartGrass® applied at 21 g/ha (0.3 oz./acre) with herbicide at unifoliate leaf (V-U) only (Fullerton) (2015)
E) RyzUp SmartGrass® applied at 21 g/ha (0.3 oz./acre) with herbicides at V-2 leaf only (Rising City) (2015)

Data collected at each site included plant heights, branching at cotyledon and unifoliate nodes, pods/plant, and yields. Data were analyzed and treatment means statistically separated using Tukey-Kramer Honestly Significant Difference test (JMP Pro 10.0.0).

Results

INTERNODE ELONGATIONS. Every GA-3 application resulted in significantly longer internodes of uppermost internode that were present at application ($P<0.0001$).

BRANCHING AT UNIFOLIATE NODE. Highly significant differences ($P<0.0001$) existed at each site for unifoliate node branching compared with untreated soybeans for GA-3 applications when initiated at unifoliate leaf (Fig. 1) but not when GA-3 was initially applied at V-1 or V-2 stages of development (Fig. 2).

As GA-3 application at unifoliate leaf stage significantly increases the distance between growing tip and unifoliate node, it is thought that auxin concentration associated with apical dominance, decreases at that node, thus allowing increased branching. Some increases in cotyledon node branching were also occasionally noted. Differences were noted between varieties for natural branching at the unifoliate node.

PODS PER PLANT. Increases in unifoliate node branching contributed to significant increases in pods/plant (Fig. 3) at all sites which received sequential applications initiated at unifoliate leaf in 2014. Differences in pods/plant were not noted in the experiments initiated at V-1 or V-2.

Harvested yields did not reflect pod counts noted in August, although large differences in pod numbers still existed at harvest. A trend was noted for yield environment however (Fig. 4). Increased pod numbers resulted in lower yield in a low fertility site (Auburn), but a large increase (12%) in a high yield environment (Shickley). This indicates that fertility/site interacted and limited yield expression as yields did not match pods at harvest at the low fertility (Auburn) location. Yield differences at any particular site were not statistically different.

GA-3 application at V-1 or V-2 increased yields compared with untreated soybeans, but sequential applications at V-1 & V-2 significantly reduced yields. As pods/plant were very similar at harvest, the reason for these yields differences is unknown but may be related to pod height above ground, thereby decreasing pods left uncombined.

Conclusion

Usage of GA-3 as a single application at V-1 or V-2 was shown to increase soybean production, although the significant increase in pods/plant from two sequential applications did not result in a consistent yield increase.

An economic analysis, using a total application cost of $42/ha ($17/acre), $31.16/100 kg ($8.50/bushel) of soybeans and an average net increase of 430 kg/ha (6.4 bushels/acre) for a single application at V-1 or V-2, indicated that the calculated economic net return value of using GA-3 at these stages of development was $92.42/ha ($37.40/acre).

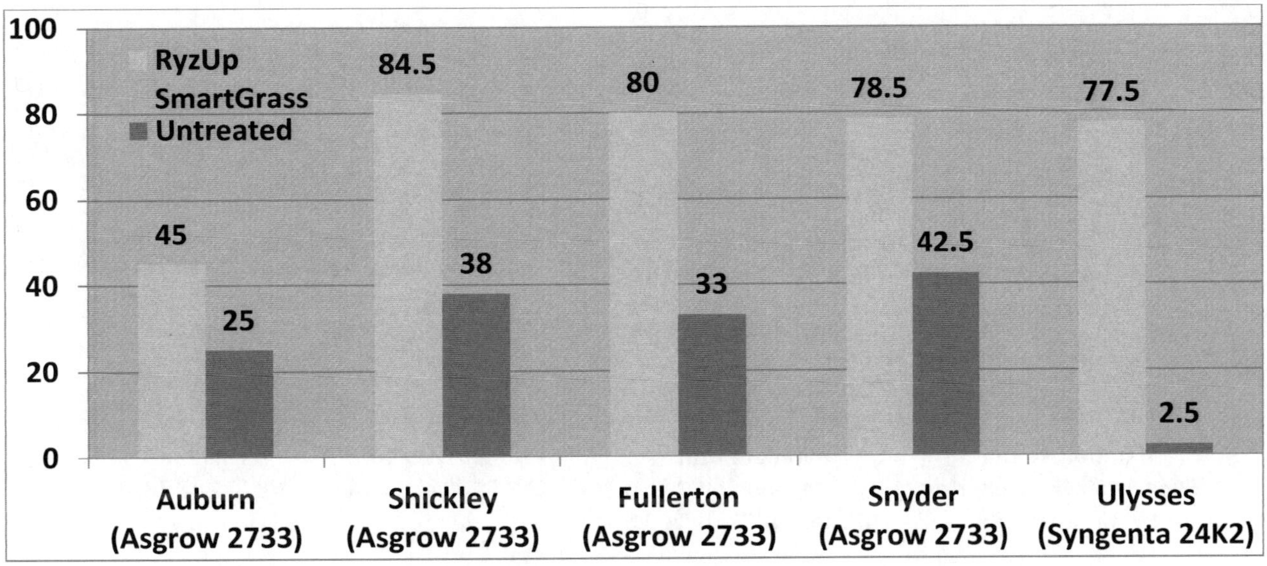

Fig. 1. Mean percent branching at soybean unifoliate nodes July 2014 following sequential applications of GA-3.

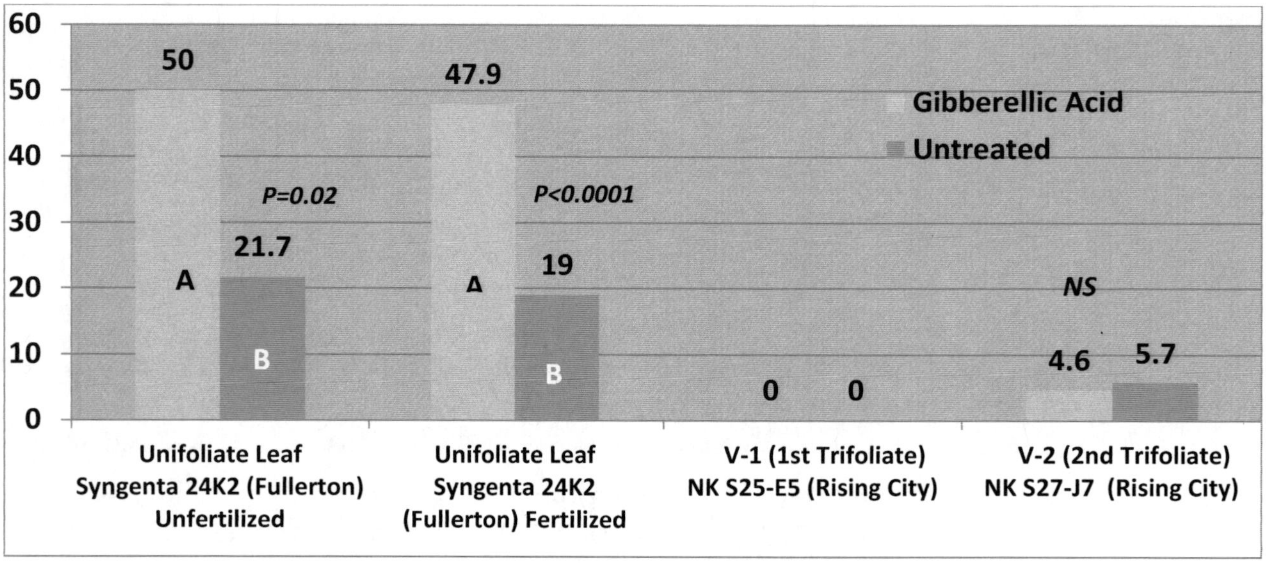

Fig. 2. Interaction of crop growth stage with a single GA-3 application on mean percent branching observed at soybean unifoliate nodes in late July.

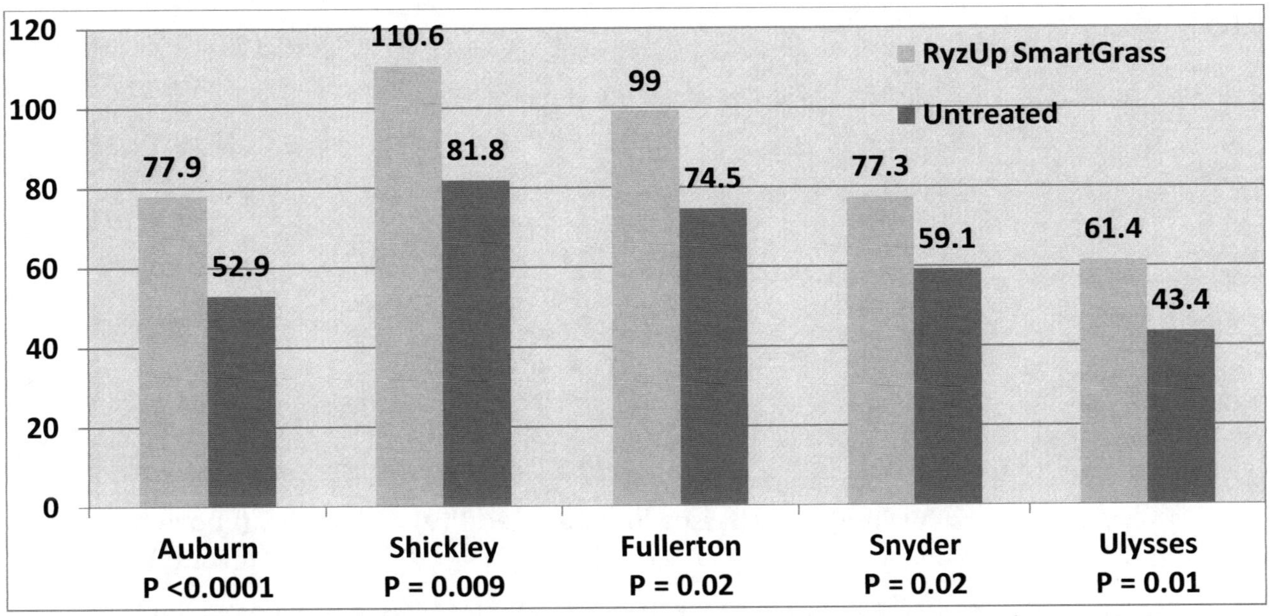

Fig. 3. Mean soybean pods per plant in early August, 2014, resulting from sequential GA-3 applications initiated at unifoliate leaf stage of development.

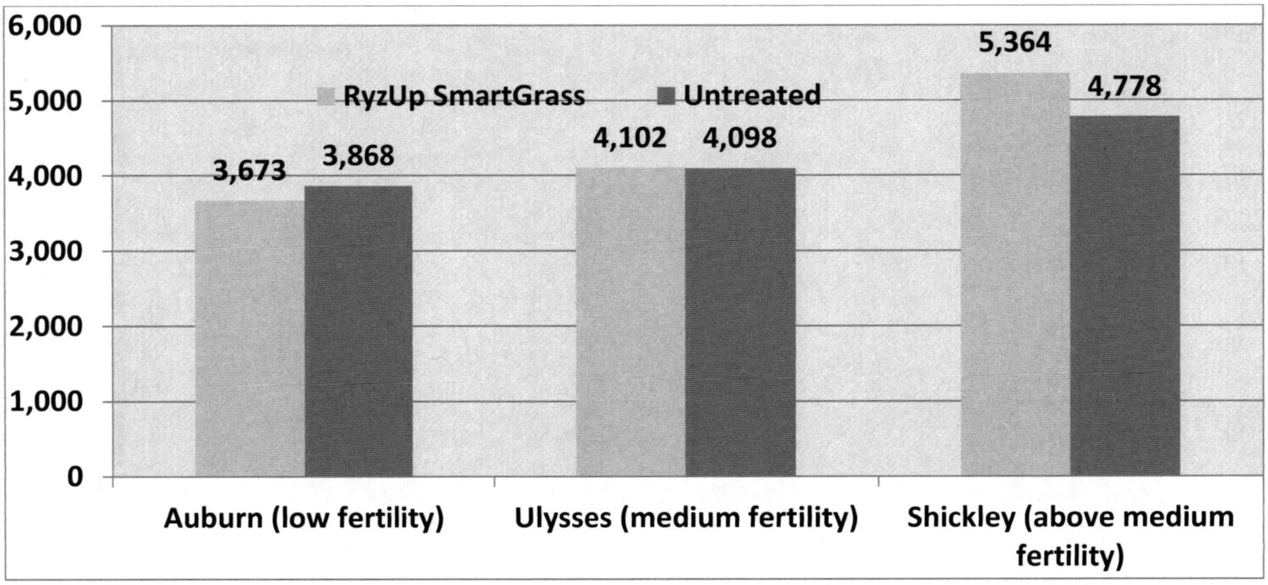

Fig. 4. Soybean yield (kg/ha) response to sequential applications of GA-3 at three levels of field fertility/yield potential.

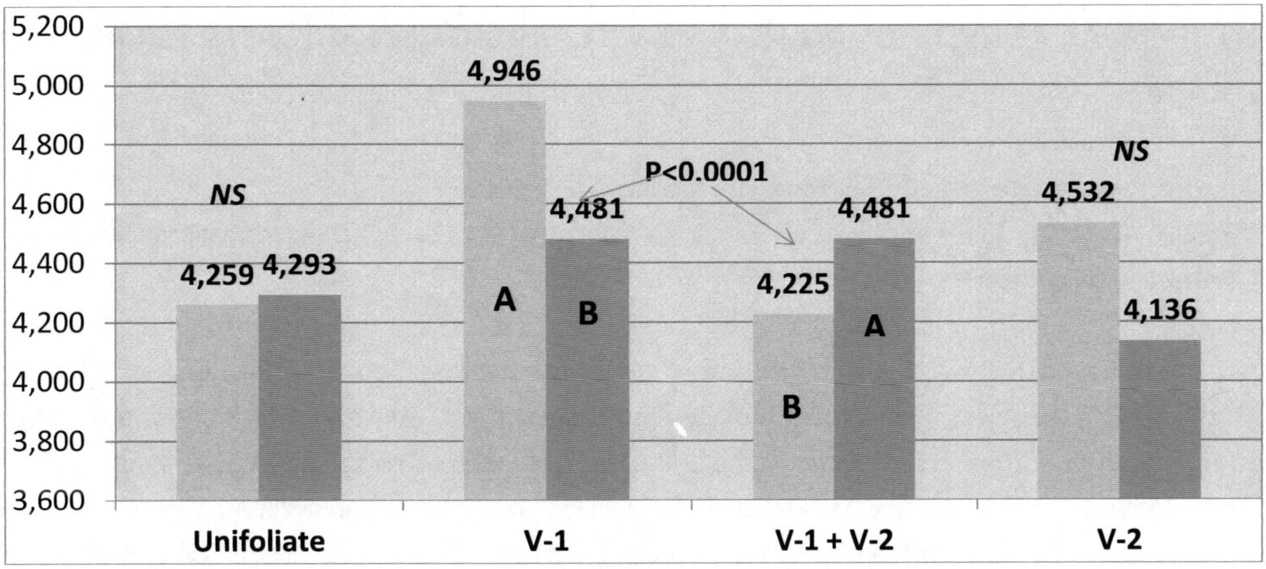

Fig. 5. Soybean Grain Yield (kg/ha) Following Single GA-3 Application at Unifoliate, V-1, or V-2 Stage, or Sequential Application at V-1 + V-2 Stages of Soybean Development.

ANTIFUNGAL ACTIVITY OF PLANT-GROWTH PROMOTING RHIZOBACTERIAL STRAINS

T. Georgieva, Y. Evstatieva, D. Nikolova, R. Tropcheva and V. Savov[1]
Sofia University "St. Kliment Ohridski", Faculty of Biology, 1164 Sofia, 8 Dragan Tsankov Blvd.
[1]Corresponding author: romb.ltd.bg@gmail.com

Plant growth promoting rhizobacteria (PGPR) are a heterogeneous group of bacteria that can be found in the rhizosphere, at root surfaces and in association with roots, which can improve the extent or quality of plant growth directly and or indirectly. In last few decades a large array of bacteria including species of genera Pseudomonas, Bacillus, Azotobacter, Klebsiella, Alcaligens, Arthobacter and Serratia have reported to enhance plant growth. The indirect promotion of plant growth occurs when PGPR lessen or prevent the deleterious effect of one or more phytopathogenic microorganisms. The most popular bacteria, studied as biocontrol agent includes the species of genera Pseudomonas and Bacillus. The aim of this work is to analyze antifungal activity of PGP rhizobacterial strains.

Antifungal activity of two strains Bacillus amyloliquefacines, three strains Bacillus subtilies and three strains Pseudomonas putida against Aspergillus flavus, Aspergilus niger, Fusarium graminearum, Penicillium claviforme and Rhizopus archisus were estimated on single layer agar method.

The two strains Bacillus amyloliquefacines are inducing larger inhibition effect compared to the other strains against all tested fungal representatives.

The strains Bacillus subtilies showed well defined antifungal activity against Aspergillus niger, Aspergillus flavus and Fusarium graminearum and lower activity against Penicillium claviforme. There was no activity against Rhizopus archisus by the tested Bacillus subtilies strains.

For any of the three Pseudomonas putida strains there was no effect of suppressing or delaying the development of the test fungi although they are good siderophore producers.

Based on the obtained data, it could be stated that two PGPR strains Bacillus amyloliquefacines demonstrated a broad spectrum antifungal activity against the five tested fungal species. The antifungal activity of the tested Pseudomonas strains did not indicate a close interaction with the production of siderophores. Further studies on the performance of these strains on the growth of plant are necessary as to uncover the mechanisms and the potential as PGPR.

SUPPLEMENTAL LED LIGHTING AS A POTENTIAL ALTERNATIVE TO CHEMICAL PLANT GROWTH REGULATORS IN ORNAMENTAL PLUG PRODUCTION

Cristian Collado[1], Mitchell A. Wrenn[1], Gregory Sheets[1], Brian E. Whipker[1], Ricardo Hernández[1], and Abhay Thosar[2]
[1]Department of Horticultural Sciences, North Carolina State University, Raleigh, NC
[2]Horticulture LED Solutions, Philips Lighting, Rosemont, IL

Plant compactness is one of the most important morphological characteristics for young plant production. Chemical growth regulators (PGRs) are commonly used to increase plant compactness; however, they are under high regulation due to their potential health risks. Light emitting diodes (LED) can be designed with a customized spectrum providing the opportunity to manipulate plant morphology using light signals. The objective of the present experiment is to compare plant growth and morphology under different supplemental lighting treatments and under PGRs in order to potentially reduce the use of PGRs in horticultural operations. Dianthus, geranium, pansy, petunia, and salvia were grown in a greenhouse from seed until the plug stage under 9.2 ± 2.4 mol·m^{-2}·d^{-1} average solar daily light integral, 22.5 ± 2.4 °C average temperature, and 46 ± 16 % relative humidity. The six growing treatments consisted of: 1) 19%blue-81%red photon flux LED supplemental lighting (19B-81R), 2) 6%blue-5%green-89%red photon flux LED supplemental lighting (6B-5G-89R), 3) High pressure sodium supplemental lighting (HPS), 4) no-supplemental lighting (control), 5) 6B-5G-89R plus Paclobutrazol (6B-5G-89R+PGR), and 5) no-supplemental lighting plus Paclobutrazol (PGR). All the supplemental lighting treatments delivered 100 µmol·m^{-2}·s^{-1} PF for 18h for a supplemental daily light integral of 6.5 mol·m^{-2}·s^{-1}. One application of Paclobutrazol (Piccolo®) was used as the chemical PGR in concentrations ranging between 2.5 to 10 ppm depending on plant species. Independent of the cultivar, plants under HPS lighting had the greatest plant height, and plants under 6B-5G-89R+PGR treatment generally had the shortest plant height. Petunia plants in the PGR treatment had 54% greater plant height than in the 19B-81R and 6B-5G-89R LED treatments. For dianthus and pansy, plants under the PGR treatment were the most compact followed by plants in the LED treatments. Geranium plants under the 6B-5G-89R LED treatment had the same plant height as those under the PGR treatment. Other plant physiological responses such as dry mass, fresh mass, leaf number, leaf area, stem diameter, anthocyanin concentration, and chlorophyll concentration were measured and will be presented.

SYNERGISTIC EFFECT OF BIOREGULATORS WITH PESTICIDES AND HERBICIDES ON IMPROVING GROWTH, YIELD QUALITY AND CROP RESISTANCE AGAINST PATHOGENS AND PESTS

S. P. Ponomarenko[1], Z. M. Hrytsaenko[2], V. A. Tsygankova[3], O.V. Babayants[4]
[1]National Enterprise Interdepartmental Science and Technology Center "Agrobiotech" of NAS and MES of Ukraine, Kiev
[2]Uman National University of Horticulture
[3]Institute of Bioorganic Chemistry and Petrochemistry, NAS of Ukraine, Kiev
[4]Selection & Genetic Institute – the National Centre of Seed-growing and Variety Study of NAAS of Ukraine, Odessa

In the greenhouse experiments the antipathogenic activity of new polycomponent bioregulators Regoplant and Stimpo has been investigated at cultivation of different varieties of winter wheat and soya bean plants on the infectious backgrounds. The high biological efficiency against phytopathogens was obtained at preseeding treatment of plants and spraying of crops in vegetation period by Regoplant (up to 98 %) and Stimpo (up to 78 %) according to control (without treatment by bioregulators). Application of bioregulator Regoplant in combination with herbicides Herbitox or Grenadier for cultivation of sugar sorghum and winter wheat showed the best indexes of plant growth and yield quality. Obtained results, may be explained by synergistic effect of bioregulator Regoplant and herbicides, as well as bioprotective effect of biostimulator Regoplant against damaging action of herbicides on plant cells. In the molecular-genetic experiments the 2nd generation of the wheat plants (infected by pathogenic micromycetes of Fusarium L. genus without treatment by bioregulators) showed the increased resistance to pathogenic micromycetes of Fusarium L. genus. Using DOT-blot hybridization method the considerable difference in the degree of homology between mRNA of control wheat seedlings and small regulatory si/miRNA of experimental seedlings (obtained from seeds of the 1st generation, infected by pathogenic micromycetes of Fusarium L. genus and treated by bioregulators) was found. It is proposed that indicated difference connected with reprogramming of plant genome under impact of bioregulators - inductors of synthesis si/miRNA with antipathogenic properties.

Introduction

The elaboration of new progressive technologies for improving the productivity and quality of products is the main task wide range of diseases caused by an abundance of pathogens and parasites (Stanton J. and Stirlin G., 1997; Ponomarenko et al., 2010) that negative impact on the genetic potential, resulting in decrease in yield and quality due to mycotoxins present in crop production (Gao X. et al., 2006). The worldwide spreading of diseases caused by parasitic nematodes leads to significant losses in the agricultural sector (Davis E.L. and Tylka G.L., 2000). For example, as a result of the violation of crop rotation regimes in North and South America at the cultivation of soybean the threshold of spreading of soybean cyst nematode Heterodera glycines was increased by 2.4 times on the 84 % plantations, leading to the loss of hundreds millions dollars (Ponomarenko S. and al., 2016).

Unfortunately, the solution of this global problem requires a lot of efforts of scientists and manufacturers, and the improving of the crop quality requires a significant use of a wide range of chemical pesticides. Without prejudice to the safety requirements according to the manufacturer's guidelines, pesticides largely impair the realization of the genetic potential of cultivated plants, violate the soil microbial biodiversity, and damage the environment and the health of farmers and consumers. The pesticides and fungicides, their compositions and analogs lead to the development of resistant strains of pathogens and reduce the genetic potential of plant varieties and their hybrid varieties (Aktar M.W.et al., 2009).

The elaboration of the alternative environmentally friendly methods, which improve plant growth and development, comes out on top, in favor of this strategy suggest materials of the First World Congress on the use of biostimulators in agriculture (Strasbourg, France, November 2012) and the Second World Congress on the use of biostimulators in agriculture (Florence, Italy, November 2015) (Tsygankova V.A., Ponomarenko, S.P. et al., 2012; Jardin P., 2015). Alternative directions include organic and integrated methods of agricultural technologies with biological control. The important interest is the strategy of induced resistance of plants using elicitors stimulating the immune system of plants, natural molecules mimicking the attack of a pathogen or danger states. The National Enterprise Interdepartmental Science and Technology Center "Agrobiotech" of NAS and MES of Ukraine jointly with the Uman National University of Horticulture conducted researches during 15 years of the more than 50 herbicides created by leading chemical companies (Bayer crop science (Germany), Du Pont International Operations (Switzerland), Nufarm GmbH & Co kG (Austria), Dow Agro Sciences GmbH (Austria), Syngenta International AG (Switzerland)) on the growth and development of the major agricultural crops. Researches were conducted with different norms of application of herbicides (40, 60, 80, 100 %) separately without plant growth regulators and with application of bioregulators created in the National Enterprise Interdepartmental Science and Technology Center "Agrobiotech" of NAS and MES of Ukraine [Ponomarenko S. et al., 2016]. In recent years, the studies of a number of August company herbicides (Russia): Fabian – for soybean growing, Herbitox – for sorghum growing, with different rates of application, both individually and jointly with Ukrainian bioregulators of third generation of "Agrobiotech" company were conducted.

The mutagenic effect of the herbicide Triflan on the germinated seeds of wheat, corn, peas was found by scientists of the Institute of Plant Physiology and Genetics, National Academy of Sciences of Ukraine in the 2001 year (Ponomarenko S. et al., 2016). It was also shown that bioregulators reduced the mutagenic effect of herbicide by 60 % and activated the oxidative-recovery systems of plants, resulting in oxidation of residues of herbicides in the crop production and in the plant rhizosphere. We found out a recovery of biodiversity in the rhizosphere: bacteria - by 40 %, micromycetes - by 32 %, and these factors have contributed to the realization of plant genetic potential (Ponomarenko S. et al., 2016). Obviously, that bioprotective action of these bioregulators is based on the synergistic effect of metabolic products (i.e. amino acids, carbohydrates, fatty acids, polysaccharides, phytohormones and microelements) of cultivated in vitro culture of fungus-endophyte of root of ginseng Panax ginseng L. and antiparasitic macrolide antibiotics - aversectines produced by soil streptomycete Streptomyces avermitilis (Ponomarenko et al., 2010).

Our molecular-genetic researches showed that these bioregulators considerably increase plant immune resistance to the different pathogens due to their stimulation action on the synthesis in the plant cells small regulatory si/miRNA that participates in RNAi-process or posttranscriptional gene silencing (PTGS) in plants (Tsygankova V.A. et al., 2014). Gene silencing, which is mediated by either degradation or translation arrest of target mRNA, has an important role in adaptive protection of plant against viruses, in genome defense against mobile DNA elements and regulation of gene expression during plant ontogenesis (Vaucheret, 2006).

Considering our previously obtained results, the great theoretical and practical interest is the possibility of application of bioregulators in combination with pesticides and herbicides in agricultural practice for decrease their toxic effect on plant cells and improving of plant growth and resistance to pathogens and pests.

The aim of our work is comparative analysis of effectiveness of application of bioregulators Regoplant and Stimpo separately or in combination with standard insecticides and herbicides Herbitox and Grenadier for improving growth, yields quality and crop resistance against various pathogens and pests.

Materials and methods

In the greenhouse experiments we compared efficiency of PGRs in combination with standard insecticides in controlling ground beetle Zabrus tenebrioides, turnip moth Scotia segetum and Chloropidae spp., and wheat nematode Anguina tritici. The experiments on comparative efficiency of bioregulators and standard insecticides were conducted in pots of 25 cm x 25 cm size; each experiment was replicated four times. Soil samples were contaminated separately by nematode, ground beetle, and turnip moth. We used winter wheat var. Dalnytska and var. Knop, and soybean var. Arcadia Odeska. Treated by bioregulators seeds of 50 plants were studied in the experiments. The bioprotective effect of bioregulators was tested according determination of the amount of damaged plantlets (in pcs/m^2) and biological efficiency (in %).

The effect of bioregulators on stability, productivity, and quality of obtained seeds, as well as on soybean resistance to pathogenic and parasitic organisms was determined. The experiments with wheat and soybean rot and mildew diseases caused by pathogens, such as Mucor spp., Rhizopus spp., Aspergillus spp., Penicillium spp., and Trichothecium roseum, were conducted on the artificial infectious backgrounds. Efficiency against pathogens was studied at low level of spores, i.e., 0.1g of spores per 1kg of seeds, and at high level of spores, i. e., 1g of spores per 1kg of seeds.
The results were compared with the effect of modern pesticides produced by leading agrochemical companies such as Alpha-Cypermethrin insecticide, Yunta Quadro insecto-fungicide, Lamardor, Selest Top, Imidacloprid, Terios and Microplant micronutrients.

We research also biological effect of bioregulator Regoplant and different doses of herbicide Grenadier and herbicide Herbitox on increase of growth and yield quality of winter wheat var. Knop and sugar sorghum var. Sylosne 42 of Ukrainian selection. It is a mid-ripening, high-performance, resistant to lodging, abscission, drought and diseases variety. This variety is recommended for cultivation in the forest steppe area, its vegetation period is 124 days. Seeds were kindly provided by the Institute of Bioenergy Crops and Sugar Beet NAASU. Cultivation of sugar sorghum var. Sylosne 42 was conducted on the experimental field of the Department of Biology of the Uman NUH, soil – the typical podzolic black soil, area – 0.73 ha. Cultivation of winter wheat var. Knop was conducted on the experimental field of the Selection & Genetic Institute – the National Centre of Seed-growing and Variety Study of NAAS of Ukraine (Odessa).

In the laboratory experiments we studied also a post-effect of bioregulators Regoplant and Stimpo on the second generation of wheat plants var. Lastivka and Princess Olga. These plants were not treated by bioregulators; however, they were obtained from seeds of the first generation of wheat plants that were infected by pathogenic micromycetes Fusarium graminearum and were treated by bioregulators (experimental plants) or not treated by bioregulators (control plants). Specificity of bioregulators post-effect was determined according to: 1) integral indexes of growth and development of control and experimental 7-day old seedlings; 2) molecular-genetic indexes: the difference in the degree of homology (according to % hybridization) between the basic constituents of plant immune system - si/miRNA isolated from experimental 7-day old seedlings and mRNA of control seedlings using DOT-blot hybridization method (Maniatis, 1982). Small regulatory si/miRNA was isolated from experimental seedlings by our elaborated method (Tsygankova, et al.,

2012).

Results

We tested bioprotective properties of bioregulators Regoplant and Stimpo at cultivation of winter wheat and soybean on the infectious backgrounds. Use of bioregulators in combination with chemical pesticides caused increase of plant resistance against different pathogens and pests. Plant growth regulators reduced phytotoxicity of chemical pesticides and stimulated immune protective reactions of plants. As a result, commercial grain yield and seed material quality were improved. The bioprotective effect of bioregulators was sufficiently high at their use on crops infected by nematodes, ground beetle, turnip moth, and chloropid flies hat (Table 1). Bioprotective effect of bioregulators did not exceed an effect achieved at the use of insecto-fungicides, e.g.,Yunta Quadro and Selest Top. However, the level of efficiency shown by Regoplant and Stimpo against wheat nematode, ground beetle, and turnip moth, and chloropid flies was just set as high taking into account their economic effect and environmental safety.

We studied also Regoplant and Stimpo bioprotective effect on soybean plants infected by dangerous pathogens of rot and mildew of soybean (Table 2). We found that bioregulators have a positive effect on growth and development of soybean seeds. They reduced the infection impact on seed development and commodity grain quality.

The results of the cultivation of sugar sorghum, var. Sylosne 42 proved the feasibility of application of bioregulator with bioprotective effect Regoplant for presowing seed treatment at the rate of 250 ml/t and joint use of herbicide Herbitox (1 l/ha) and Regoplant (50 ml/ha). The most significant result is obtained by presowing seed treatment with Regoplant and crop spraying with Regoplant and Herbitox (Tables 3-5).

The analogical results were obtained at the cultivation of winter wheat var. Knop, which was treated during germination and vegetation period by bioregulator Regoplant and herbicide Grenadier (Fig. 1).

Our laboratory experiments demonstrated the inheritance of wheat plants resistance to pathogenic organisms. We found that plants of the 2nd generation which were not treated with bioregulators, maintain high viability which is similar to the results obtained on the 1st generation of plants treated with bioregulators on infectious background. The molecular-genetic analysis using DOT-blot method hybridization si/miRNA with mRNA populations showed high degree of homology between immune-protective small regulatory si/miRNA and mRNA of experimental plants and lower degree of homology according to mRNA of control plants (Table 6). We called this effect "quasi-heterosis". It was found that Regoplant and Stimpo strongly increased growth rate of heterosis plants as well as resistance to pathogenic organisms. We concluded that principal mechanism of these bioregulators in plant cells includes almost twofold increasing of the synthesis (abundance) of small regulatory si/miRNA, which has antipathogenic properties.

Discussion

It is known that plant growth is the result of the different processes such as cell differentiation and development, and formation of tissue, organ and whole organism; these basic processes are controlled by genetic program of plant growth and development (Tsygankova V.A., 2015). The plant cell endogenous system of coordination and self-regulation includes a variety of endogenous regulatory systems, among which an important role belongs to physiologically active compounds such as phytohormones (Adam G., 2012; Tsygankova V.A., 2015). It is not surprising, taking account diverse impact of phytohormones on plant growth and development, improving of yield and product quality and increase of plant resistance to unfavorable environmental factors (Pieterse C.M.J. et al., 2012; Denancé N. et al., 2013).

Today there are thousands of new synthesized compounds, although only a few from them are used in the agricultural practice as plant growth regulators (PGRs) (Gianfagna T.J., 1987; Basra A.S., 2000). Even such classes of recently identified substances as brassinosteroids (BR) which also belong to phytohormones are not widely used because their manufacture is not organized [Adam G., 2012] The worldwide use of PGRs is restricted due to insufficient knowledge of the mechanisms of their regulatory action, taking into account the biological characteristics of plants and high selectivity of action of PGRs that have different structure. Unfortunately, the empirical approach is used often to the selection and application of growth regulators on various plant crops, without regard biological characteristics of each plant and without consideration of the changes in metabolic processes that occur in plant cells under PGRs influence, as well as the dependence of PGRs effects on plant growth from the environmental factors and so on.

Our long-term and multiple studies show that exogenous PGRs of different nature have a significant influence on many metabolic processes in plant cells due to their impact on the genetic and phytohormonal levels of regulation, however, depending on genotype of the rate of change of these processes is different (Tsygankova, Ponomarenko et al., 2011). Considering the results of our previous research and discussion in the literature about the nature of the relationship between exogenous PGRs with the endogenous regulatory system, our attention for the last years was paid to the search for new growth regulators that possess the ability to the direct changes in plant growth and development, increasing their resistance to unfavorable factors environment.

In the present work the bioprotective effect of new bioregulators Regoplant and Stimpo has been investigated at cultivation of different varieties of winter and soya plants on infectious backgrounds. It was found that preseeding treatment of plants and spraying of crops in vegetation period by Regoplant and Stimpo increased plant viability and resistance to different phytopathogens. The highest bioprotective activity against phytopathogens was obtained at preseeding treatment of plants by Regoplant (up to 98 %) and Stimpo (up to 89 %) bioregulators according to control (without treatment by bioregulators).

Comparative study of effectiveness of using of bioregulator Regoplant separately or in combination with herbicide Herbitox for cultivation of sugar sorghum showed that the best results were obtained at the presowing seed treatment with Regoplant and crop spraying with Regoplant and Herbitox. The analogical results were obtained at the cultivation of winter wheat var. Knop, which was treated during germination and vegetation period by bioregulator Regoplant and herbicide Grenadier. Obviously, these facts are as a result of the synergistic effect of bioregulator Regoplant and herbicide Herbitox or herbicide Grenadier, as well as bioprotective effect Regoplant on decrease of toxic effect of herbicides on plant cells. This technology can be recommended to farmers as Ukrainian innovative project.

In the laboratory experiments it was shown that the seedlings of the second generation of wheat plants (that were grown on infectious background without treatment by bioregulators), obtained from the seed of the first generation of plants (that were grown on infectious background and were treated by bioregulators Regoplant and Stimpo), show high viability and resistance to pathogenic micromycetes of Fusarium L. genus due to integral indexes of growth and development comparatively with control seedlings. The obtained molecular-genetic indexes testify that at the phase forming of seeds there is observed the reprogramming genome of plant embryos under influence of bioregulators through the way of activation of expression of genes encoding antipathogenic si/miRNA.

Literature Cited

Aktar M.W., Sengupta D. and Chowdhury A. Impact of pesticides use in agriculture: their benefits and hazards. Interdiscip Toxicol. 2009. 2(1): 1–12.

Adam G., Duke S.O., Gross D., Lischewski M., Marquardt V., Vaughn K.C., Voigt B. Herbicide Resistance — Brassinosteroids, Gibberellins, Plant Growth Regulators. Chemistry of Plant Protection (Volume 7). W. Ebing (ed). Springer Berlin Heidelberg, 2012. 176 p.

Ponomarenko S. et al., 2016. Ukraine plant growth regulators from idea to reality, p. 165-167. Conference 27-28.04.2016 Ukraine.

Basra A.S. (Ed). Plant Growth Regulators in Agriculture and Horticulture: Their Role and

Commercial Uses. The Haworth Press. Inc., New York. 2000. 264 p.

Ciancio A. and Mukerji K.G. (eds). Integrated management and biocontrol of vegetable and grain crops nematodes. Volume 2, 2008. 356 p.

Denancé N., Sánchez-Vallet A., Goffner D. and Molina A. Disease resistance or growth: the role of plant hormones in balancing immune responses and fitness costs. Frontiers in Plant Science. Plant Cell Biology. 2013. 4 (Article 155): 1–12.

Davis E.L. and Tylka G.L. Soybean cyst nematode disease. The Plant Health Instructor. 2000.

Maniatis T., Fritsch E. F., Sambrook J. Molecular cloning: A laboratory manual. New York: Cold Spring Harbor Lab, 1982. 480 p.

Gao X., Jackson T.A., Hartman G.L., Niblack T.L. Interactions Between the Soybean Cyst Nematode and Fusarium solani f. sp. glycines Based on Greenhouse Factorial Experiments. Phytopathology. 2006. 96(12): 1409-15.

Jardin P. Plant biostimulants: Definition, concept, main categories and regulation. 2015. 96 (30): 3–14.

Ponomarenko S. al., Plant growth regulators in agrobiocenosis: new solutions. In "Plant physiology in Ukraine at the edge of the millennium", 2001, p. 145-151.

Mycotoxins: Risks in Plant, Animal and Human Systems, Task Force Report / Council for Agricultural Science and Technology, Ames, IA 2003.

Ponomarenko S.P., Terek O.I., Grytsaenko Z.M. Bioregulation of plant growth and development. – In S.P. Ponomarenko, H.O. Iutynska (eds). Bioregulation of microbial-plant systems. Nichlava Press, Kyiv, 2010. P. 251 – 291.

Pieterse C.M.J., Van der Does D., Zamioudis C., Leon-Reyes A., and Van Wees S.C.M. Hormonal Modulation of Plant Immunity. Annual Review of Cell and Developmental Biology, 2012. 28: 489-521.

Tsygankova V. A., Ponomarenko S.P., Galkin A. P., Galkina L. O. and Eakin D. Gene expression under regulators' stimulation of plant growth and development. In S.P. Ponomarenko, H.O. Iutynska (eds). New plant growth regulators: basic research and technologies of application. Nichlava Press, Kyiv, 2011. P. 94 – 152.

Tsygankova V.A., Andrusevich Ya.V., Ponomarenko S.P., Galkin A.P., and Blume Ya.B. Isolation and Amplification of cDNA from the Conserved Region of the Nematode Heterodera schachtii 8H07 Gene with a Close Similarity to Its Homolog in Rape Plants. Cytology and Genetics. 2012. 46(6): 335-341.

Tsygankova V.A., Ponomarenko, S.P., Hrytsaenko, Z.M. Increase of plant resistance to diseases, pests and stresses with new biostimulants, Acta Horticulturae, Strasburg (France). 2012. 1009: 225–233.

Tsygankova V.A., Iutynska G.A., Galkin A.P., Blume Ya.B., Impact of New Natural Biostimulants on Increasing Synthesis in Plant Cells of Small Regulatory si/miRNA with High Anti-

Nematodic Activity. Int. J. Biol. 2014. 6(1): 48-64

Tsygankova V.A. Genetic Control and Phytohormonal Regulation of Plant Embryogenesis. Int. J. Med. Biotechnol. Genetics (IJMBG). 2015. 3(1): 9 - 20.

Stanton J. and Stirlin G., Nematodes as plant parasites, In: Plant Pathogens and Plant Diseases, Edited by J.F.Brown and H.J.Ogle. Armidale, Australia: Rockvale Publications, 1997. P. 127-142.

Vaucheret H. Post-transcriptional small RNA pathways in plants: mechanisms and regulations. Genes and Devel. 2006. 20, 7: 759 - 771.

Table 1. Bioprotective effect of bioregulators and seed protectants against phytopathogens.[z]

Experiment	Application rate (L/t)	Wheat nematode		Ground beetle		Turnip moth		Chloropid flies	
		Amount of infected plants	Biological efficiency (%)	Amount of infected plants	Biological efficiency (%)	Amount of infected plants	Biological efficiency (%)	Amount of infected plants	Biological efficiency (%)
Control		42.5		36.6		15.2		39.4	
Stimpo	0.025	22.5	47	14.6	60	9.9	35	17.1	57
Regoplant	0.25	5.4	87	6.8	81	5.9	61	10.1	74
Yunta Quadro	0.15	3.5	92	0.1	100	0	100	2.1	95
Selest Top	0.2	4.1	90	1.9	95	0	100	2.8	93
Imidacloprid	1.0	17.9	58	1.1	97	0.6	96	2.1	95
Alpha-cypermethrin	0.5	29.9	30	9.6	74	4.5	70	7.1	82
LSD 0.05[y]		1.1		0.9		0.8		2.3	

[z]All experiments were replicated four times. Total amount of plants in each experiment was 50.

[y]LSD 0.05 – Least significant difference

Table 2. Bioregulators efficiency against the causative agents of rot and mildew of soybean.[z]

Experiment	Application dose (L/t)	Fusarium sp.		Botrytis cynerea		Alternaria spp.		Complex of storage fungi[y]	
		Amount of infected seeds	Bioeffiency (%)	Amount of infected seeds	Bioeffiency (%)	Amount of infected seeds	Bioeffiency (%)	Amount of infected seeds	Bioeffiency (%)
Control		43.6		29.3		9.7		32.4	
Stimpo	0.025	12.8	71	8.1	72	1.1	89	8.9	73
Regoplant	0.25	1.8	96	0.6	98	0.5	95	4.5	86
Lamardor	0.2	1.5	96	0	100	0	100	0	100
Yunta Quadro	1.5	0	100	0	100	0	100	0	100
LSD 0.05		0.8		0.7		0.6		1.4	

[z]All experiments were replicated four times. Total amount of plants in each experiment was 50.

[y]Fungi Mucor spp., Rhizopus spp., Penicillium spp., Trichothecium roseum.

Table 3. The dynamics of leaf surface area during the application of different doses of herbicide Herbitox and different ways of use of growth regulator Regoplant.

	Tillering		Ear emergence		Milky-wax ripeness		Content of chlorophylls	
Variant	Leaf area on one plant (cm^2)	%	Leaf area on one plant (cm^2)	%	Leaf area on one plant (cm^2)	%	Sum of chlorophylls a and b	%
Control	44.1	100.0	367.6	100.0	426.5	100.0	3.719	100.0
Herbitox 1.0 L/ha	51.6	117.0	385.9	105.0	464.9	109.0	3.89	104.6
Background + Herbitox 1.0 L/ha + Regoplant 50 mL/ha	64.6	146.5	460.6	125.3	542.5	127.2	5.106	137.3
LSD 0.05	1.78		3.29		4.42		4.42	

Table 4. Weed infestation.

Variant	No. of weeds	Weight of weeds	Killing of weeds (%)	
	Pcs./m^2	g/m^2	By quantity	By weight
Control	32	135.1	0	0
Herbitox 1.0 L/ha	13.4	83.9	58.1	37.9
Background + Herbitox 1.0 L/ha + Regoplant 50 mL/ha	4.3	16.5	86.6	87.8
LSD 0.05	0.78	-	-	-

Table 5. Impact of growth bioregulator Regoplant and herbicide Herbitox on yield of sorghum green mass and grain.

Variant	Yield of green mass		Grain yield		
	t/ha	%	t/ha	± to control	%
Control	32.46	100.0	3.52	-	-
Herbitox 1.0 L/ha	33.18	102.2	3.86	+0.34	+9.7
Herbitox 1.0 L/ha + Regoplant 50 mL/ha	34.77	107.1	4.14	+0.62	+17.6
Background + Herbitox 1.0 L/ha + Regoplant 50 mL/ha Full technology	38.73	119.3	4.60	+1.08	+30.7
LSD 0.05	0.313	-	0.187		-

Table 6. The degree of hybridization between si/miRNA with mRNA of wheat plants. [z]

Plant	Cultivar	Hybridization (%)		
		Control	Regoplant	Stimpo
Wheat	Lastivka	98 ± 1.4	82 ± 1.6 (≈16%)	86 ± 1.2 (≈12%)
	Princess Olga	98 ± 1.6	84 ± 1.4 (≈14%)	88 ± 0.96 (≈10%)

[z]All experiments were conducted in triplicate. Significant differences from control values, $P < 0.05$.

Fig. 1. Impact of the bioprotective plant growth regulator Regoplant (ISTC Agrobiotech) and herbicide Grenadier (China) on the yield of winter wheat (var. Knop) in comparison with the different systems of basic cultivation. Institute of Agriculture (Odessa)

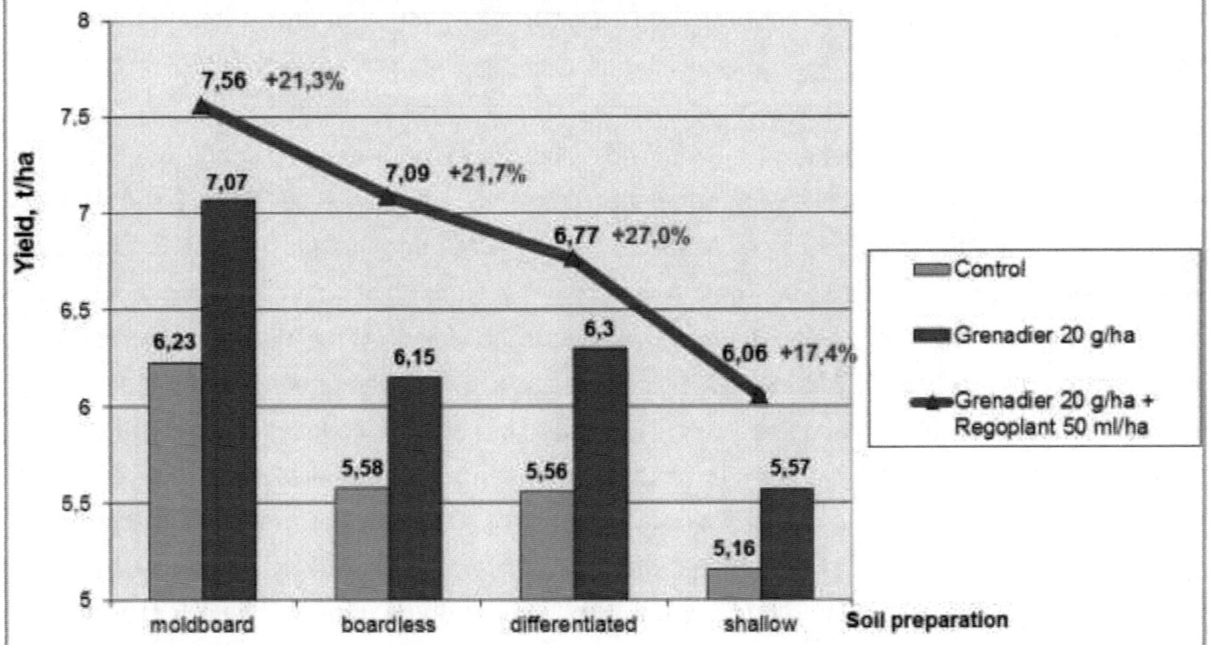

HEALTH BENEFITS OF ABSCISIC ACID IN GLYCEMIC CONTROL

Raquel Hontecillas[1,2], Josep Bassaganya-Riera[1,2], Adria Carbo[1] and Noah Philipson[1]
[1]Biotherapeutics Inc., 1800 Kraft Dr. Suite 200, Blacksburg, Virginia 24060
[2]Nutritional Immunology and Molecular Medicine Laboratory (www.nimml.org), Virginia Tech, Blacksburg, VA 24060

Pre-diabetes, metabolic syndrome (MS) and diabetes are serious health conditions that increasingly afflict the majority of the U.S. population. Lanthionine synthetase C-like 2 (LANCL2) has been identified as the molecular target of abscisic acid (ABA) for glycemic control. Plasma and tissue ABA levels originate from two sources: 1) dietary origin from plants, and 2) endogenous synthesis. The plasma ABA pool can be depleted in diabetic individuals. Oral ABA administration in microgram amounts significantly reduces both glycemia and insulinemia of healthy subjects undergoing an oral glucose tolerance test (OGTT), indicating that ABA reduces the amount of insulin required to control hyperglycemia. To further characterize the anti-diabetic effects of low-doses of ABA we used the db/db and the diet-induced obesity (DIO) mouse models of diabetes. In dose response studies, db/db mice were treated with oral concentrations of 1, 10, 40 and 200μg ABA/kg/d. Oral ABA treatment (1μg/kg/d) of genetically obese db/db mice and mice with DIO significantly reduced fasting blood glucose (FBG), glucose tolerance during a GTT, and HbA1C levels. Furthermore, we provide mechanistic evidence in vivo that the molecular target of ABA is LANCL2 since the loss of the receptor in LANCL2-/- mice abrogated the glucose-normalizing effects of oral ABA. Mechanistically, ABA increased translocation of the glucose transporter GLUT4 to the plasma membrane and GLUT4-dependent glucose uptake in adipocytes and myoblasts. Release from entero-endocrine cells of GLP-1, a key regulator of carbohydrate intake and metabolism targeted by the most recent anti-diabetic drugs, is enhanced by ABA both in vitro and in vivo. In summary, oral ABA supplementation is emerging as an innovative nutritional intervention for improving glycemic control in healthy individuals and pre-diabetic patients by activating LANCL2.

TENDRIL PRODUCTION IN CROSSES OF CROWN PEA WITH A DETERMINATE LINE OF PEA

Sonja L. Maki
Department of Plant and Earth Science, 410 South Third Street, University of Wisconsin-River Falls, River Falls, WI 54022

Interest has been gaining in the development of hyper tendriled and semi-leafless lines of pea with increased tendril production for the gourmet vegetable market. The current research explored whether a fasciated mutation could affect characteristics such as tendril production. Crown Pea (Fig. 1), a determinate, fasciated line of pea in a tall background (Le) was crossed with a determinate, cochleata line in a dwarf background (le) that occasionally makes extra leaflets to determine the affects on plant growth and development.

Alterations in leaf and axillary bud structure was observed in the F2 generation (Figs. 2 and 3). Leaflets from F2 plants occasionally produced tendriled leaf tips (Fig, 3). Axillary buds often did not develop or were replaced by leaf-like structures (Figure 3). The leaf-like axillary bud phenotype continued in the F3 generation, however some individuals did develop axillary buds occasionally. In addition, more anthocyanin production was observed in leaflet tissue of the segregating for the cochleata phenotype (Fig. 4).

Plant height, final number of nodes, average internode length, and stipule phenotype was measured for nine F2 plants (Table 1). Differences in plant height, average internode length and stipule phenotype were observed in the F2 generation.

Future research will focus on further characterization of the segregates generated in this study and in progeny from similar crosses with semi-leafless varieties as well as the effects of gibberellin application and light quality on tendril production.

Table 1. Plant characteristics of nine F2 plants from the crown pea x det coch cross.

Plant	Mature Height (cm)	No. of Nodes	Avg. Internode Length (cm)	Stipule Phenotype
1	96	20	4.8	wildtype
2	85	19	4.5	wildtype
3	60	17	3.5	wildtype
4	32	18	1.8	cochleata Nodes 7-9 simple Nodes 10 and > compound*
5	23	15	1.5	wildtype
6	24	17	1.4	cochleata all simple
7	77	20	3.85	cochleata all simple
8	100	19	5.3	wt
9	75	19	3.9	cochleata Nodes 7-15 simple, Nodes 16 and > compound*

*no stipules developed on these plants below node 7.

Fig. 1. Crown pea at flowering.

Fig. 2. F2 of crown pea x det coch cross at flowering displaying a determinate flowering habit.

Fig. 3. F2 of crown pea x det coch cross later in development displaying tendriled leaf tips and leaf-like axillary structures.

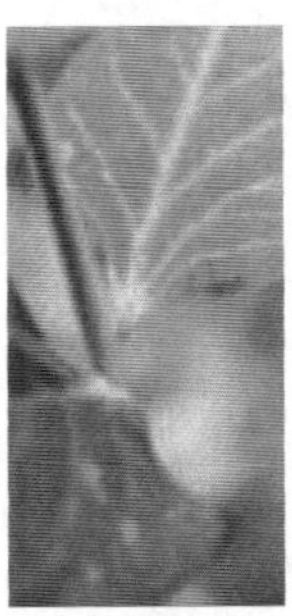

Fig. 4. Aberrant anthocyanin production was also observed in the leaflets of F2 progeny in the crown pea.

BIOSTIMULANTS: USING MODERN ANALYTICAL APPROACHES TO CHARACTERIZE AND QUANTIFY COMPLEX BIOACTIVE INGREDIENTS

David Hiltz
Acadian SeaPlants

Biostimulants are generally regarded as products that enhance the growth, hardiness, and productivity of plants above and beyond the effect of its nutrient content. This term has been used for the last two decades to describe a variety of substances, mostly derived from natural material, including humic and fulvic acids, aquatic plant extracts, protein hydrolysates, and microbial products. As one can see, the nature of these various classes of biostimulants is very diverse, as are the effects on the physiology and biochemistry of the plants to which they are applied. In many cases, these products are not single compounds and are instead complex mixtures of a variety of bioactive components of biological origin that have not been fully characterized. There is significant research showing the intended agricultural function of biostimulants, but linking the effects to specific compounds can be challenging. Attempts to characterize biostimulants have included classical natural products isolation techniques along with analytical tools and most recently, metabolomics and genomic approaches. Here we present an overview of some historical and recent approaches to characterizing and identifying the various classes of biostimulants in an attempt to gain further understanding of their potential modes of action.

ROLE OF ADJUVANTS ON SPRAY DROPLET SPECTRA OF PLANT GROWTH REGULATORS

Marcus Jones, Jason Haegele, Jon Zuk, Lillian Magidow and Andrea Clark,
WinField Solutions

Plant growth regulators (PGRs) are routinely used to manipulate growth and development of horticulture and agricultural crops. They elicit plant responses at low concentrations and have a narrow optimum dose response range. PGRs may be applied to the soil or as a foliar spray and are often applied in conjunction with other crop protection products or mineral nutrition. The effectiveness of foliar applications is dependent upon successful delivery to the target where the spray solution may be absorbed by the plant tissue. Non-uniform distribution of the prescribed dose over the intended target could affect PGR efficacy and lead to inconsistent performance. This study evaluated the droplet spectra variety of PGRs alone and in combination with glyphosate as affected by adjuvants using a Sympatec HELOS-KR laser diffraction particle size sensor in a fully enclosed low speed wind tunnel. Active formulations were delivered using field-appropriate rates, nozzles, and pressures in a fully enclosed low speed wind tunnel. Air flow of 8 ± 0.5 mph was concurrent with the spray direction, preventing repeated sampling of very fine particles. The effect of adjuvants on various spray parameters was measured. These parameters included: (1) the cumulative percent of spray volume comprising droplets smaller than 105 μm, or "driftable fines," and the percent change in this quantity relative to a standard; (2) the volume median diameter (DV0.5), or diameter at which 50% of the volume includes smaller droplets, as well as the DV0.1 and DV0.9 (10% and 90%, respectively); and (3) the relative span, calculated using (DV0.9-DV0.1/DV0.5).

Plant Growth and Regulation Society of America
1018 Duke Street
Alexandria, VA 22314
USA

ISBN 978-1-5108-4499-5

9 781510 844995

CURRAN ASSOCIATES INC.
proceedings.com